T0292549

Ascidians in Coastal Water

H. Abdul Jaffar Ali · M. Tamilselvi

Ascidians in Coastal Water

A Comprehensive Inventory of Ascidian
Fauna from the Indian Coast

 Springer

H. Abdul Jaffar Ali
Department of Biotechnology
Islamiah College (Autonomous)
Vaniyambadi, Tamil Nadu
India

M. Tamilselvi
Department of Zoology
V.V. Vanniaperumal College for Women
Virudhunagar, Tamil Nadu
India

ISBN 978-3-319-29117-8 ISBN 978-3-319-29118-5 (eBook)
DOI 10.1007/978-3-319-29118-5

Library of Congress Control Number: 2015960399

Printed on acid-free paper

This Springer imprint is published by SpringerNature
The registered company is Springer International Publishing AG Switzerland

Foreword

India is one among the top ten species-rich nations of the world blessed with high degree of endemic species. It has four global biodiversity hot spots namely Eastern Himalaya, Indo-Burma, Western Ghats and Sri Lanka and Sundaland among the 32 of the world. It also has various ecozones such as deserts, high mountains, highlands, tropical and temperate forests, swamp lands, plains, grass lands, areas surrounding rivers, as well as coastal and marine habitats besides island archipelagos. Among these the coastal and marine habitats are important as India has a vast coast length of about 8100 km and has biodiversity rich habitats such as estuaries, lagoons, backwaters, mangroves, salt marshes, rocky coasts, sandy beaches and dunes. India also has an Exclusive Economic Zone (EEZ) spreading over 2.02 million km^2 (0.86 million km^2 on the west coast, 0.56 million km^2 in the east coast and 0.6 million km^2 around the Andaman and Nicobar islands). The coral reefs present in the coastal areas teem with abundant life. All the above habitats are having a unique biodiversity characterizing each habitat.

Among the countries in Asia, India has perhaps a unique distinction of having a long record of inventorying the coastal and marine biodiversity going back at least two centuries. While commercially important organisms and their taxa received considerable importance in faunal investigations, other group of organisms in particular the invertebrates and prochordates received only cursory attention. One such group is the Ascidians belonging to subphylum Prochordata. This phylum has two subphyla namely Cephalochordata and Urochordata. Globally a total of 3061 species has been reported under this phylum. From Indian waters 464 species of prochordates under 63 genera and 18 families have been reported. The subphylum Urochordata includes class Ascidiacea (sea squirts), class Thaliacea (salps) and the class Larvacea (Doliolum etc.). Classes Thaliacea and Larvacea include members which are planktonic in nature while the ascidiacea includes sessile and benthic organisms attached to the substratum of coral reefs etc. Among the three classes, class Ascidiacea is the most speciose one. As many as 400 ascidian species have been known from India. However this group has not been fully explored and much remains to be done. Also the known species have not been taxonomically described. In addition there is no consolidated information, a publication, a monograph or compendium having the list of species, their description, and notes on their distribution and ecology which can motivate other researchers to concentrate on this group of organisms. There is a growing need for the above as ascidians with their anti-bacterial, anti-tumour, anti-fouling and insecticidal properties have become important in the fields of biotechnology, biochemistry, microbiology, biomedicine, immunology, pharmacology and drug discovery. They also serve as sentinel organisms-bio-indicators with their ability to accumulate metals. They also constitute a non-conventional food source. In view of the wide application values of this group of organisms, the above publication has become highly imperative. Having been working in this group of organisms for more than a decade, **Dr. H. Abdul Jaffar Ali and Dr. M. Tamilselvi** felt the existing lacuna in Indian Ascidian research and have come out with this important publication. This book provides a comprehensive overview of ascidians and also covers the taxonomic characteristics of a total of 58 species of ascidians belonging to 20 genera and 9 families collected from the southern coasts of India. The photographs add value to this publication. This book will be immensely useful and indispensable to those who plan further works on this group of organisms besides researchers, teachers, students, administrators and planners. The authors of this book deserve deep appreciation for the serious efforts they have taken to bring out this compilation. Their Mentor **Dr. V. Sivakumar** deserves all the praise for motivating, inspiring and guiding them. I am confident that this book will be received well by those interested in ascidian research.

Parangipettai
October 2015

S. Ajmal Khan
Professor Emeritus
Faculty of Marine Science
Centre of Advanced Study in Marine Biology

Preface

There is a popular belief among the biologists that ascidians are rare and extremely poor with respect to number of species and abundance. In spite of being conspicuous and macroscopic, very little attention has been paid to this group in India. Even though there is a wealth of ascidians in India, active research is lagging not only in the taxonomical point of the ascidians but also on other aspects such as biochemistry, embryology, evolution, microbiology, toxicology, breeding, and aquaculture. In recent times, a lack of appreciation for taxonomic work is discernible in scientific circles leading to concentrate on areas other than systematics. However, new concepts of the conservation and management of biodiversity and related problems have clearly revealed the great relevance and significance of taxonomy. This book provides a comprehensive overview of ascidians and also covers the entire taxonomic characteristics of the collected species from southern coasts of India with relevant photographs. The presented data and taxonomy would form the basis for future monitoring, leading to bioinvasion control, and also generate new arena of research on ascidians. Helpful suggestions and constructive criticism for the improvement are always welcome.

Acknowledgements

We would like to thank the **Almighty** for his blessings showered on us and for leading us into the path of enlightenment. The authors wish to express their grateful thanks to **Dr**. **V**. **Sivakumar**, Director of Research and Conservation, 4e India NGO (Reg 188/2010), our chief mentor, for his contribution, constructive suggestions, and guidance throughout our research period. Thanks are also due to Mr. L.M. Muneer Ahmed, B.Sc., Secretary, and Dr. K. Prem Nazeer, Principal of Islamiah College (Autonomous), Vaniyambadi, for their generous supports and providing a great research environment. This research was financially supported by the Department of Biotechnology, Government of India, Grant Number BT/PR6801/AAQ/3/609/2012 to Dr. H. Abdul Jaffar Ali. We continue to be thankful to our research scholars, Mr. A. Soban Akram, M.Sc., and Mr. M.L. Mohd Kaleem Arshan, M.Sc., for their tremendous efforts in preparing the manuscript and drawings. We would like to thank the publishers especially for providing us an opportunity to publish a book on Ascidians that faces a dearth of quality publications. Special recognition goes to our family members for their love, patience, and understanding during all long hours of work.

H. Abdul Jaffar Ali
M. Tamilselvi

Contents

Preamble

'The Ocean' is a mysterious paradise in nature, housing billions of species many of which are life-saving species. Conducive environment of the ocean provides plenty of food, shelter, and everything forever for their sound survival. Being a self-sustaining and self-regulating system, ocean maintains the structure and functions of the marine ecosystem. The diversity of marine life underpins all biological studies. Many animals in the ocean live and vanish even without being noticed by us. Hence, they hardly find a place in the faunistic list of our natural resources. The biological diversity of each country is a valuable and vulnerable natural resource. Sampling, identifying, and studying biological specimens are among the first steps towards protecting and benefiting from biodiversity. It is well understood that every species makes some contribution to the structure and function of its ecosystem. In this context, Ascidians, belonging to the subphylum Urochordata, are gaining paramount importance as they contribute a lot to the stability of the marine ecosystem by providing a fertile ground for a number of aquatic fauna, a part of food chain, prey for many marine animals, store house of bioactive compounds and serve as indicators to assess the quality of water. They are also fished and farmed in many parts of the world for food particularly in Korea, France, Japan, and Chile.

Ascidians form the ubiquitous portion of benthic communities in shallow tropical and temperate waters. Yet, the ascidian fauna of many regions of India is still poorly surveyed and the identification of species by non-specialists almost nonexistent. Most ecological studies of sessile communities include quite a few species, but frequently published lists include identification only to family or genus levels. The lack of field guides and identification keys pave the ways to unaware of biodiversity in the specific area which is directly related to economic growth of a nation. In the present scenario, it is an inevitable task to provide incessant supply of food to the geometric increase of human population in the world. Therefore, the socioeconomic and ecological development of the country is mainly depends on the 'biodiversity'—the biological wealth and health of the ecosystem. According to UN, marine life provides protein-rich food for over 2.6 million people worldwide and offers potential biomedical compounds to treat various kinds of diseases. Even though the utility of the acknowledged species has already been reached its maximum extent due to overfishing, habitat destruction by natural calamities, and anthropogenic effects, certain species remains in the place of 'endangered' list, in turn the mankind to seek alternative source of food from the marine environment. Ascidians are one such species that can fulfill the demand of future generation due to their more availability and adaptability to sustain in the marine environment.

Abstract

Ascidian research in India only began in the second half of the twentieth century. Currently about 400 ascidian species are known in India, most of them are not taxonomically described. Recent increased interest in pharmacological research on ascidians throughout the world has highlighted the lack of knowledge available on ascidian species, distribution, abundance, and diversity in India. The lack of basic information on Indian ascidians is a great concern. Due to ever-changing coastal diversity and topographic nature of the collection spots, a special attention has been made on the ascidians since ecological knowledge such as diversity and distribution is in incipient. The present account consists of two parts: Part I which comprises 'General Account on Ascidians' covering all aspects of ascidians in brief and Part II which gives a 'Comprehensive Inventory of Ascidian Fauna of Southern Indian Coasts.'

The sampling for ascidians was done from 11 coastal stations along southwest and southeast coastline of India focusing in natural and artificial substrates at various marine habitats. The taxonomic coverage of this data set spans the class Ascidiacea and included the updated knowledge on the spatial distribution of ascidians along the southern coastline of India. A total of 58 species of ascidians belonging to 9 Families and 20 Genera were recorded. All the species were described with photographs and diagrams. Eight out of a total species treated in this work were new record to Indian water. Many species were invasive, some were established invasive, and few were cryptogenic. The range of forms treated in this work was largely the result of efforts of hand picking and snorkeling method of collection.

The most diverse genera in these waters were *Eudistoma* and *Polyclinum* and well represented in southeast coast of India. Of the families, Styelidae and Didemnidae were well represented. The largest number of records was from the family Didemnidae and Polycitoridae (N = 13), followed by Polyclinidae (N = 12), Styelidae (N = 6), Pyuridae (N = 5), Ascidiidae (N = 4), Perophoridae (N = 3), and Rhodosomatidae and Molgulidae (N = 1).

The presented data and taxonomy would form the basis for future monitoring, leading to bioinvasion control, and to the generation of new arena of research on ascidians. In addition, ascidian has a great potential to be used as a candidate species for novel compounds, alternate protein-rich source for animals as well as human beings, bioindicator of water pollution, etc. This content will be definitely very helpful for researchers, coastal planners, port authorities, coastal thermal plants, and atomic power plants for proper management of these ecologically significant ascidians.

Keywords Ascidian · Diversity of ascidians · Taxonomy of ascidians · Distribution of ascidians · Indian ascidians · Inventory of ascidians · Invasive ascidians

Part I
General Account on Ascidians

Chapter 1
An Introduction to Ascidians

1.1 What Are Ascidians?

Ascidians are exclusively marine prochordates belonging to the Class Ascidiacea of the subphylum Urochordata. There are both solitary and colonial forms with a sessile adult attached directly to hard substrates or fixed in sediments and a tailed, free-swimming larva.

Ascidians go by several names.

1.1.1 Tunicate

The outer delicate body "mantle" is enclosed in a hard or transparent jacket like protective structure is known as "test" or "tunic" hence so called tunicates. The tunic is made up of cellulose like material, tunicin which is a very rare occurrence within the animal kingdom.

1.1.2 Sea Squirt

When disturbed or once the animals are removed from the water, they eject small amount of water through branchial (or) feed sac/siphons, hence so name by "sea-squirt".

© Springer International Publishing Switzerland 2016
H.A. Jaffar Ali and M. Tamilselvi, *Ascidians in Coastal Water*,
DOI 10.1007/978-3-319-29118-5_1

1.2   Relationship with Vertebrates

Despite their simple appearance and very different adult forms, their closeness towards the vertebrate is shown by the presence of notochord and tadpole like appearance during their motile larval stage. The name tunicate derives from unique outer covering called "tunic" which is formed from proteins and carbohydrates and acts as an exoskeleton. In some species, it is thin, translucent, and gelatinous, while in others it is thick, tough, and stiff. Tunicates are the members of the sub-phylum Urochordata, closely related to the phylum Chordata that includes all vertebrates. Because of the close ties with Chordates, many scientists are paying keen attention to understand about their biochemistry, developmental biology, and genetic relationship with other invertebrate and vertebrate animals. There are many evidences to show that tunicates are related to vertebrates: tunicate tadpole larva has a nerve cord down its back, similar to the nerve cord found inside the vertebrae of all vertebrates including human beings. The cerebral vesicle is equivalent to a vertebrate's brain. Sensory organs include an eyespot, to detect light, and an otolith, which helps the animal orient to the pull of gravity. After the larval stage the notochord and dorsal nerve cord are absent in the adults. Tunicates are classified in the subphylum Urochordata, signifying that the notochord is present only in the tail of the larva in fact, they are the only sessile chordate.

1.3   Characteristic Features of Ascidians

The important characteristic features of ascidians are

- Ascidians are exclusively marine.
- They are solitary or colonial forms.
- Adult leads sessile or sedentary mode of life but the larval forms are free swimming.
- The body is covered by an outer protective "test" or "tunic".
- Presence of larval notochord at the tail region prompted the inclusion under Urochordata.
- They exhibit primitive mode of feeding such as filter, ciliary and mucous.
- Presence of wide pharynx with numerous gill slits in adult resembles the chordate characters.
- Presence of notochord in a confined region of the tail of larva. Hence it is included in the subphylum Urochordata.
- Presence of specialized cells in blood called vanodocytes (exception-*Herdmania momus*).
- Absence of definite excretory system.
- Notochord is reduced to small ganglion in the adult.
- Most tunicates are hermaphrodites.
- Reproduction is by sexual and asexual methods.

- Development is indirect.
- Existence of a free swimming "tadpole larva" in the life history trait.

1.4   Types of Ascidians

Based on the morphological features, ascidians are classified into three categories and are simple/solitary, colonial/compound and social/synascidian (Fig. 1.1).

1.4.1   Solitary Ascidians

Simple ascidians are solitary in nature and enclosed by the test in sole form and attached to the substratum by its foot. They are single and discrete individuals that can only reproduce sexually.

1.4.2   Colonial Ascidians

Whereas in colony forming ascidians, all the zooids are enclosed in a common test and all zooids may or may not connected with basal stolons, opens to the exterior directly or through a common cloacal aperture. Naturally, in its specific environment, it is a spacial competitor for other sedentary invertebrates such as sponges and barnacles.

Practically, it is very difficult to identify the colonial tunicates from sponges during field collection. The gelatinous nature of the former one and rough surface of the latter (owing to the presence of spicules) helps to identify these organisms in field visit.

1.4.3   Social Ascidians or Synascidians

They are genetically-identical individuals that are vascularly connected to each other in some way. Each "zooid" or discrete unit has an incurrent and excurrent siphon.

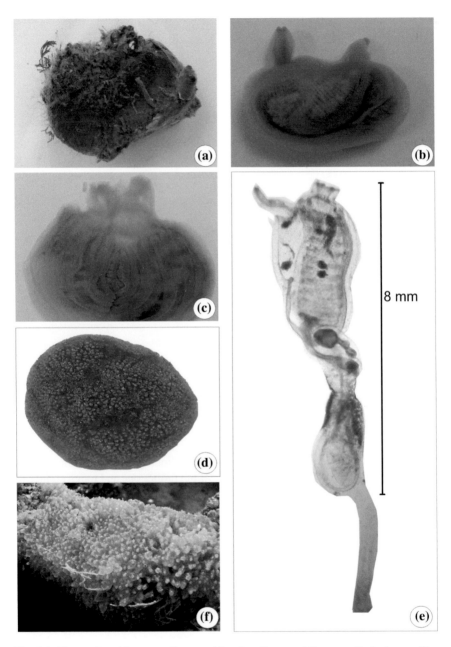

Fig. 1.1 Types of ascidians **a** solitary ascidian **b** solitary ascidian—mantle body **c** solitary ascidian—mantle body opened **d** colonial ascidian—colony **e** colonial ascidian—zooid **f** synascidian

1.5   Vernacular Names of Various Ascidians

Dark Ascidian:	*Botrylloides leachi*
The Red-throated Ascidian:	*Herdmania momus*
Giant Pink Ascidian:	*Botryllus* sp.
Orange-throated Ascidian:	*Stomozoa australiensis*
Blue-throated Ascidian:	*Clavelina molluccensis*
Diminutive Ascidian:	*Pycnoclavella diminuta*
Little Orange Ascidian:	*Stolonica australis*
Brain Ascidian:	*Sycozoa cerebriformis*
Red-mouthed Ascidian:	*Herdmania grandis*
Mauve-mouth Ascidian:	*Polycarpa viridis*
Black Ascidian:	*Phallusia nigra*
Club Ascidian:	*Polycarpa clavata*
Southern Sea Tulip:	*Pyura australis*
Lumpy Sea Tulip:	*Pyura gibbosa*
Southern Ascidian:	*Sigillina australis*
Cyan Ascidian:	*Sigillina cyanea*
Red Crater Ascidian:	*Aplidium clivosum*
Gelatinous Ascidian:	*Cystodytes dellachiajei*
Grape Ascidian:	*Clavelina cylindrica*
Pink Encrusting Ascidian:	*Didemnum ancanum*
Puffball Ascidian:	*Aplidium brevilarvacium*
Noddy Ascidian:	*Sycozoa pedunculata*
Murrays Ascidian:	*Sycozoa murrayi*
Cushion Ascidian:	*Neodistoma* sp
White Glove leather:	*Didemnum albidum*
Yellow or Obese Ascidian:	*Phallusia obesa*

1.6   General Biology of Ascidians

1.6.1   Feeding Mechanism

As ascidian leads a sedentary mode of life, it depends on a continuous uninterrupted flow of water for getting food as well as oxygen and also to eliminate the excreta. Ascidian exhibits three types of feeding mechanisms such as filter, ciliary and mucous feeding for getting food from the massive environment. Constant beating of cilia in the stigmata creates continuous water current which brings the food particles such as diatoms, algae, protozoans etc. Majority of the ascidians are plankton feeders, some are detritivorous which live on soft sediments (*Molgula* sp.). Very few

act as sit—and—wait predators inhabiting deep water region and trap small invertebrates such as tiny crustacean and nematodes by using strong muscular lobes
around the buccal sipons. Some ascidians (*Didemnum* sp.,) living in tropical areas
show symbiotic relationship with green algae or cyanobacteria present in their
tunics. Besides, the efficiency of feeding rate depends on the rate of water current
generates and the amount of food particles enter into the pharynx. Water enters into
the branchial or oral siphon and flows into a wide branchial chamber crossing the
mucous ciliated gill slits and then into a water chamber called atrium and finally
expelled out throught the atrial siphon. During this process, cilia in the stigmata help
to prevent the escape of planktons through the outgoing water. Large sized food
particles are also retained by the tentacles of the oral siphon at entry level. The
filtered food particles are entered into the wide pharynx and are retained in the
branchial sac where they are entangled with the mucus, secreted by the glandcells of
the endostyle. The entangled food cord is rolled into a cylindrical mass and transported into the dorsal lamina. In the dorsal lamina, after the conversion of cylindrical
mass into a string is carried into the oesophagus and from where it reaches the
stomach. Digestion occurs in the stomach and intestine by the enzymes secreted
from the liver and pyloric glands respectively. The digested food is absorbed in the
intestine, and undigested food is eliminated through the atrial siphon.

1.6.2 Respiratory System

Unlike other vertebrates, there are no well defined structures for respiration. The
wide branchial sac or pharynx is the major site of respiration and inner surface of
the mantle wall also provides the secondary site for gaseous exchange. Exchange of
oxygen and carbon dioxide between the blood and the water medium occurs by
diffusion.

1.6.3 Circulatory System

The circulatory system of ascidian is peculiar and unique. They have a well
developed heart, blood vessels and specialized corpuscles in the blood. The double
U shaped heart is tubular in nature and enclosed in a pericaridum and connecting at
either end to a system of blood vessels viz; 1. Ventral aorta from the ventral end of
the heart runs on the ventral side of the pharynx, 2. Dorsal aorta which is found on
the dorsal side of the pharynx and 3. Branchio-visceral vessel arises from the
posterior end of dorsal aorta and starts from the dorsal end of the heart and divides
into small branches and ends in the oesophagus, stomach intestine, liver, gonad and
test. The heart and blood vessels have no valves to control the pumping of blood.
The blood is pumped around the body by a series of peristaltic contraction starts
from one end of the heart to another. The blood contains plasma and corpuscles.
Tunicate blood contains pigments in the vanodocytes which is an unusual feature in

ascidian. Hence, certain members of Ascidiidae and Perophoridae have the ability to accumulate high concentration of the transitional metal vanadium and vanadium associated proteins in vacuoles of blood cells known as vanodocytes. Certain species belonging to the family Pyuridae can accumulate iron due to the presence of tunichromes in the blood cells. Hence the colour of the blood is red in *Herdmania momus* or lustering yellowish green in *Phallusia nigra*.

The mechanism of circulatory system adds this group in a special place in the animal kingdom. Due to the presence of two pace makers in the heart of the ascidians, a reversal of blood flow takes place in the heart and blood vessels i.e. every few minutes, the heart stops beating and then restarts makes the pumping fluid flows in the reverse direction. As a result, the heart pumps oxygenated blood (systemic heart) at one time and deoxygenated blood (branchial heart) at other time.

1.6.4   Excretory System

Ascidians do not have any form of excretory organs such as kidney or a nephridial system to excrete nitrogenous waste materials. Ascidians exhibit ammonotelism and uricotelism type of excretion. There are some cells known as nephrocytes do the job of excretion when they are loaded with products like xanthine and urate particles. Some species such as *Ciona intestinalis*, *Ascidiella aspersa* and *Molgula manhattensis* excrete ammonia as excretory product. In Molgulidae, a large vesicle is situated adjacent to the heart which contains partly composed of uric acid. The absence of uricolytic enzyme prevents the uric acid from further degradation to urea or allantoin.

1.6.5   Nervous System

The nervous system is poorly developed in adult tunicates but in the larva, there is a well developed dorsal tubular nerve cord with brain and sense organs. In the adult, there is a single, solid, long ganglion called the brain situated in the dorsal mantle wall mid-way between the branchial and atrial siphons and adjacent to the neural gland. The nerve ganglion gives off three and two nerves towards the branchial and atrial siphon respectively. There are no sense organs but sensory cells are present in the siphons, buccal tentacles and in the atrium.

1.6.6   Reproduction

Different reproductive strategies are obviously adopted by different species of ascidians to produce genetical and non-genetical individuals by asexual and sexual reproduction respectively. Most of the colonial ascidians reproduce asexually by budding and fragmentation. Budding helps to establish the species in its local origin whereas the fragmentation facilitates reproduction as well as dispersal of the species. Budding is a natural process in which zooids replicate within a colony, thereby the colony is grown and new colonies are also produced. Budding of colony enables the species to survive in unfavourable environmental conditions. It is remarkable to note that, in Diazonidae and Polycitoridae the epidermis cuts through the abdomen to form buds, whereas in Polyclinidae, the epidermal strobilation is through the post-abdomen. In Didemnidae, a pair of buds is formed by outpushing from the limbs of the abdomen, in which, the anterior bud develops into an abdomen and posterior into thorax. In Perophoridae, the buds arise from the ectodermal ventral vessel, which originates from the base of the endostyle and grows into a great length. In *Botryllus schlosseri*, budding occurs in all seasons although the process is slower in winter. The colonial ascidians with very thick and fleshy colonies show fragmentation very rare and never reattached (*Aplidium constellatum*). The sessile and lobed colonies readily produce fragments. Once fragments are produced, they may be transported to the new sites by water currents, waves and tides and they reattach to the substratum (e.g. *Botryllus schlosseri* and *Didemnum* sp.). When an individual colony dies after fragmentation, its potential immortal genotype may survive in other colonies.

Generally, solitary ascidians reproduce sexually by releasing the gametes into atrium through the gonoducts and discharged into the sea water where fetilization takes place. The spawning period varies from species to species. Eg. *Ciona intestinalis* and *Molgula manhattensis* release their eggs and sperm 1–1.5 h before sunrise and *Corella parallelogramma* and *Styela partita* do so in the late afternoon. In general, the hermaphroditic gonads release matured ova earlier than the sperms. This condition is called as protogynous that helps to avoid self fertilization. In this case, the released eggs into the seawater are fertilised by sperm from another individual in the seawater. As fertilization takes place in the water, it is said to be external fertilization. In most solitary ascidians, the fertilization and development takes place in sea water but almost all compound ascidians retain their eggs within the oviduct, atrial cavity, brood pouch or in the test matrix until the larva is complete and able to swim. In colonial ascidian, fertilization is internal. Generally, the egg is fertilized in the abdomen of the zooids and passes into the test directly from the abdomen and liberated as swimming larva known as "tadpole" resembling the frog's tadpole larva.

1.6.7 Ascidian Tadpole Larva

It is very interesting to note that the behavioural pattern of the larva and the adult are quite different from one another in many respective features. The larva shows positive phototrophism and negative geotrophism whereas the adult shows negative phototrophism and positive geotrophism. During the larval period, it does not take any feed as the mouth is closed by the test. The larval period varies from a few minutes to several days. After a brief planktonic life, the larva becomes attached to some objects using the adhesive papillae at the anterior end. With regards to functional significance for the attachment, the structures such as adhesive papillae, anterior ampullae and epidermal vesicels show structural variations in some species of ascidians tadpole larva. For illustration, in the members of the family Polyclinidae, the presence of different kinds of papilla, including deformed papilla force out the secretary cells to secrete adhesions which facilitate the attachment of larva with the available substrates. During the larval life, it retains some important chordate characteristic features; a notochord is located throughout the length of the tail but is absent from the trunk. Above the notochord, a tubular nerve cord is found and its anterior end extends into the trunk.

1.6.8   Metamorphosis

After a brief planktonic life, the larva becomes sluggish and sinks to the bottom where it attaches itself by adhesive papillae and stands erect but with head downwards and then undergoes metamorphoses. During this event, the structures important to lead a successful larval life are degenerated and new structures are developed to lead the remaining life during adult. In other words, more advanced larva changes into a less developed adult which is called retrogressive metamorphosis. The following structures such as tail, notochord, nerve cord, muscles and fins are completely lost. The trunk ganglion degenerates and the posterior part of the cerebral vesicle changes into a ganglion. The region between the mouth and adhesive papilla grows rapidly whereas the opposite region becomes degenerated resulting in the development of branchial and atrial apertures by rotation of its body at an angle of 180°. Later, pharyngeal stigmata increase in number and the adult has well developed heart. The test increases in thickness which helps to strongly anchor the substratum to lead a successful adult life in the stressed environment.

Chapter 2
Classification of Ascidians

The mysterious morphological structure of ascidians was taken into account for classification by various researchers. Herdman (1882) has given a classification of tunicate based upon the external features in relation to gemmation and formation of colonies. Originally the suborders are designated as the orders Aplousobranchia, Phlebobranchia and Stolidobranchia (Lahille 1886) based upon the complexity of the branchial sac. Subsequently, Perrier (1898) devised the orders Enterogona and Pleurogona based upon the position of the gonads and other morphological considerations. Garstang (1896) put forward a scheme of classification based on anatomical and embryological characters and included, to a certain extent, Lahille's system of classification. Garstang (1928) and Huus (1937) combined the classification of Lahille and Perrier incorporating Lahille's Aplousobranchia and Phlebobranchia as suborders of the Enterogona with the Pleurogona containing only the suborder Stolidobranchia (Berril 1950). Storer and Usinger (1965) classified the sub phylum Urochodata and divided it into three classes namely Ascidiacea, Larvacea and Thaliacea. In most modern treatments the class is divided into the orders Enterogona and Pleurogona with the sub orders Aplousobranchia and Phlebobranchia in the Enterogona and a single suborder, the Stolidobranchia in the Pleurogona (Abbott et al. 1997; Kott 1985). In recent classification (Kott 2003), tunicata comprises into four classes. They are:

Class I—Ascidiacea
The class Ascidiacea includes sedentary ascidians. There is a free-swimming larval period in the life history of an ascidian. Individuals are simple/solitary or compound/colonial in nature. Both siphons (branchial and atrial) are pointed upward in direction. Atriopore is dorsal in nature. Pharynx is perforated with numerous gill-slits opens into the atrium. Dorsal lamina may or may not found in the endostylar region. Adult has no tail and notochord. Sexual reproduction occurs. Eg. *Herdmania momus*, *Phallusia arabica* and *Didemnum psammathodes*.

© Springer International Publishing Switzerland 2016
H.A. Jaffar Ali and M. Tamilselvi, *Ascidians in Coastal Water*,
DOI 10.1007/978-3-319-29118-5_2

Class II—Larvacea (Appendicularia)

This group is also known as appendicularia due to the presence of appendage (tail) used for propulsion. Larvacea includes small, transparent, planktonic group of tunicate with temporary test. Larvaceans are solitary in nature. These animals live in a gelatinous casing or "houses" and retain their larval tail throughout their lives. This tail drives a gentle current of water through the house, propelling the organism in the water. Pharynx has only two gill slits, opens to the outside of the body as there is no atrium. They are hermaphrodites except *Oikopleura dioica*. This group retains neoteny. Eg. *Oikopleura vanhoeffeni*.

Class III—Thaliacea

Members of the class Thaliacea never attach to objects, but lives as planktonic drifters. Thaliaceans are strange gelatinous animals that use their siphons to jet-propel themselves gently through the water. Mouth and atripore are on opposite ends providing propel for locomotion. Transparent test may or may not well develop. Life history includes alternation of generation. They occur from the ocean surface down to around 1500 m. Eg. *Pyrosoma atlanticum*, *Salpa* and *Doliolum* sp.

Class IV— Sorberacea

The Sorberacea is a class of benthic ascidian that has a dorsal nerve cord in the adult stage; in contrast to the sea squirt. They are carnivorous in habit. There is no perforated branchial sac. They are carnivorous in habit. Example: *Sorbera* sp. , *Gasterascidia* sp. , *Oligotrema* sp. , *and Hexadactylus* sp. ,

2.1   Class: Ascidiacea

Most recently, the search of ascidians have increasingly and become the target for many research activities, particularly on the members of class Ascidiacea for its double mode of traits in their life history and distinguished from other classes by its sessile habits. So far, more than 3000 species of ascidians are recorded from all over the world. Among the four classes, more number of families and species are recorded from the Class Ascidiacea. The individuals of the class Ascidiacea comes under two orders. They are Enterogona and Pleurogona.

Order 1: Enterogona

Members of this order comprise both simple and colonial ascidians. The body of the simple form is undivided, whereas in colonial forms the body divides into thorax and abdomen. Tentacles are simple. Unpaired gonads are situated on one side or behind the intestinal loop. Neural gland is present ventral to neural ganglion. Oviduct and sperm duct follow the rectum and open through anus. Larvae are having cerebral eye and otolith.

Order Enterogona consists of two suborders:

Suborder 1: Aplousobranchiata Lahille, 1886

(=Haplobranchia Garstang, 1895, and = Krikobranchia Seeliger 1906).

It comprises only colonial ascidians. Zooids are elongated and divided into distinct thorax and abdomen. Post abdomen may or may not be present. Branchial sac has no folds and inner longitudinal vessels, simple, transverse, ciliated ridge is present. Gut loop is present in abdomen. Gonads are present in the abdomen or post abdomen.

There are three families included in Aplousobranchiata.

Family 1: Clavelinidae Forbes and Hanley, 1848
It includes colonial forms with most of them are social ascidians. Zooids are more or less free or embedded in matrix. Body is divided into thorax and abdomen. Gill slits ranged from three to many. Atrial siphon opens to extensive directly or into a common cloaca. True post abdomen is absent. Gonads are situated within the loop of intestine. This family comprises three subfamilies.

Subfamily 1: Polycitorinae
Zooids open to exterior independently with atrial siphon. Both atrial and branchial siphon are 6 lobed. Asexual reproduction is common and occurs by budding of abdominal constriction

Genera: *Polycitor, Eudistoma, Archidistoma, Tetrazoma, Cystodytes* and *Sigillina.*

Subfamily 2: Clavelininae
Solitary and colonial forms. The siphons are smooth bordered and opens to the exterior directly. Internal longitudinal branchial vessels are absent.

Genera: *Clavelina, Podoclavelina, Pycnoclavelina, Archiascidia, Oxycorynia.*

Subfamily 3: Holozoinae
Zooids are arranged in systems. Branchial siphon has six lobed, whereas atrial siphons are the modified atrial languets. Brood pouch with embryos present at thoracic level. Buds are produced at end of the abdomen.

Genera: *Distaplia, Sycozoa, (Corella), Holozoa.*

Family 2: Polyclinidae Milne-Edwards, 1842 (=Synoicidae Hartmeyer, 1908)
It includes only colonial ascidians, usually massive and gelatinous. Zooids are large with seven or more rows of stigmata. Branchial aperture is lobed, whereas atrial aperture is not lobed. Atrial siphon opens into common cloaca. Body of the zooid is divided into thorax, abdomen, posterior abdomen. Gonads and heart are present in post abdomen. Budding is by post abdominal constrictions. The family Polyclinidae includes two subfamilies.

Subfamily 1: Polyclininae
Polyclinids with atrial apertures forming common cloaca, branchial siphon six or eight lobed.

(a) Branchial siphon six lobed, stomach smooth-walled.
    Genera: *Polyclinum, Aplidiopsis, Sydneioides.*
(b) Branchial siphon six lobed, stomach with ridges.
    Genera: *Aplidium, Amaroeucium, Synoicum.*

(c) Branchial siphon eight lobed, stomach with ridges.
   Genera: *Sydnyum, Morchellium.*
(d) Aberrant genus. *Pharyngodictyon.*

Subfamily 2: Euherdmaniae
Polyclinid with free atrial siphons. Epicardia separate.
   Genus: *Euherdmania.*

Family 3: Didemnidae Giard, 1872
Exclusively colonial forms. Colonies are encrusted and the zooids are very short
divided into thorax and abdomen. Atrial siphons are modified to form atrial
lip. Calcareous spicules are present in the test. Gonads are present dorsal or posterior. Budding is complex involving epicardia.
   Genera: *Didemnum, Trididemnum, Diplosoma, Leptoclinides, Lissoclinum,
Echinoclinum, Leptoclinum, Coelocarmus.*

Suborder 2: Phlebobranchiata Lahille, 1885
(=Dictyobranchia Seeliger, 1906).
   Both solitary and colonial forms. Body is not divided into thorax and abdomen
in Diazonidae. Branchial sac is with internal longitudinal vessels or in the form of
bifurcating branchial papillae, but never with folded. Gonads are present in gut loop
or at the side of branchial sac. Phlebobranchiata includes seven families.

Family 1: Cionidae Lahille, 1887
Solitary in habit. The digestive tube is horizontal and posterior to thorax. Branchial
sac is large contains many rows of stigmata. Inner longitudinal vessels with sec-
ondary papillae. Heart is V-shaped situated between stomach and base of endostyle.
Gonads are present in gut loop, both oviduct and spermduct open near the anus.
Generally oviparous.
   Genus: *Ciona.*

Family 2: Diazonidae Garstang, 1891
Both solitary and colonial forms. Each zooid has separate atrial and branchial
siphons. Body is divided into a thorax and an abdomen, gut loop is vertical
enclosing gonads. The branchial sac with numerous rows of stigmata, inner longitudinal vessels are present with primary papillae. Secondary papillae are absent.
Gonad is in gut-loop with ducts opening near the anus.
   Genera: *Diazona, Tylobranchion, Rhopalea.*

Family 3: Perophoridae Giard, 1872
Exclusively colonial and individuals are separate from one another. Gut loop is
lower and at the side of branchial sac. The branchial sac with longitudinal vessel is
in the form of bifurcating primary papillae. Secondary papillae are absent. The heart
is long up to the base of endostyle along the dorsal side. Epicardia are absent.
Gonad is in gut-loop; sperm duct opens near the anus. Whereas oviduct is very
short and opens into brood pouch chamber in matrix cavity.
   Genera: *Perophora, Ecteinascidia.*

Family 4: Corellidae Lahille, 1887 (= Rhodosomatidae Hartmeyer, 1908)
Exclusively solitary. Gut loop is partly below, partly at the right side of the branchial sac. The branchial sac has inner longitudional vessels and without secondary papillae. Epicardia are present. Gonads in gut-loop and both ducts open near the anus.

Genera: *Corella, Rhodosoma, Chelyosoma*, and some imperfectly known abyssal genera.

Family 5: Ascidiidae Herdman, 1880
Exclusively solitary forms. The gut-loop is entirely on the left side of the branchial sac. The branchial sac has inner longitudinal vessels, with or without secondary papillae. The epicardia are represented by numerous renal vesicles. Gonads are in gut loop and both ducts open near the anus.

Genera: *Ascidia, Ascidiella, Phallusia*.

Family 6 and 7: *Hypobythiidae* Sluiter, 1895, and *Agnesiidae* Huntsman, 1912
Contain respectively one and two highly specialized deep-water genera probably related to *Ascidiidae*.

Order 2: Pleurogona
Both simple and colony forming ascidians. The body is never divided into the thorax and abdomen. Gonads are in the lateral mantle wall on both sides. The neural gland is usually dorsal, sometimes lateral to ganglion. Tentacles may be simple or compound. Buds are produced from the mantle wall.

Suborder 1: Stolidobranchiata Lahille, 1886. (=*Ptychobranchia* Seeliger, 1906)
Solitary or colony forming pleurogonid ascidians. Branchial sac with inner longitudinal vessels or bars, usually with longitudinal folds. This suborder comprises three families.

Family 1: Styelidae Sluiter, 1895
Both solitary and colony forming ascidians. Siphons are smooth edged or fringed with four lobes. Tentacles are simple. Branchial sac is always 4 folds on each sides. Stigmata straight. Dorsal lamina has smooth-edged. Stomach is wide ridges and pyloric caecum.

Subfamily 1: Styelinae
Solitary styelids: Both siphons have four-lobed. Usually four folds on each side of the branchial sac. Gonads are on one or both sides, usually multiple.

Genera: *Pelonia, Styela, Katatropa, Polycarpa, Cnemidocarpa*.
Gonads on one side only.
Genera: *Dendrodoa, Podostyela*.

Subfamily 2: Botryllinae
Colony forming styelids; colonies are either compact or with zooids more or less free. Buds are produced from the mantle wall. Gonads are very variable. Tadpole larvae are with peculiar sensory "photolith".

(a) Branchial sac with folds and numerous inner longitudinal vessels or bars.
     Genera: *Polyandrocarpa, Eusynstyela, Gynandrocarpa, Stolonica, Distomus.*
(b) Branchial sac without folds, but with residual longitudinal vessels or bars.
     Genera: *Botryllus, Botrylloides, Symplegma, Metandrocarpa, Kukenthalia, Dictyostyela, Chorizocarpa, Polyzoa.*

Family 2: Pyuridae Hartmeyer, 1909

Exclusively solitary ascidians. Both siphons are four lobed. Tentacles are branched. Branchial sac is above with four folds on each sides. Many internal longitudinal vessels are present. Stigmata are straight. The stomach is narrow and smooth.
     Genera: *Pyura, Boltenia, Microcosmus, Tethyum, Culeolus.*

Family 3: *Molgulidae* Lacaze-Duthiers, 1877

Exclusively simple ascidians with both siphons are six lobed. Tentacles are branched. The branchial sac has 6 or 7 branchial folds with numerous longitudinal vessels. Stigmata are curved and arranged in spirals. The renal organ is a large closed sac lying on left side.
     Genera: *Molgula, Eugyra.*

Chapter 3
Ecology of Ascidians

3.1 Are Ascidians "Key Stone Species"?

Ascidians/tunicates are one of the dominant members in the benthic macrofauna and contribute more significantly to raise the biodiversity of marine ecosystem by extending their distribution on a variety of natural as well as the artificial substratum across the world. Also, they inhabit diverse marine habitats such as intertidal rocky areas, muddy areas, sandy beach, sea grass beds, coral reefs, etc., available from the tropics to the poles and from shallow water to the deep sea, even from hadal region. Scientists from all over the world have shown keen interest on tunicates as their recruitment, dispersal, survival and reproduction are dependent on environmental variables, which in turn, they govern the existence and extension of other groups of organism. This trait of leading life raised a number of questions in connection with the evolutionary point of view as this group adopted the principles of "The survival of the fittest" in the theory of natural selection of Charles Darwin for the survival during earlier and later life by possessing preadaptation and post adaptation respectively. Inspite of living under stressed environment, they provide tranquil environment for a number of organisms. They contribute a part in food chain, act as storehouse of potential primary and secondary metabolites, offer provision of food and medicines, prey for number of organisms, serve as indicators to assess the quality of water, alarm the status of entry of non-native species and their succession. Therefore, ascidians are considered as keystone species in the benthic megafaunal community, as their loss from a conservation area leads to the loss of numerous other species as inadvertently reserved other species. The change in ascidian composition echoes the changes in environmental variables as well as community composition.

© Springer International Publishing Switzerland 2016
H.A. Jaffar Ali and M. Tamilselvi, *Ascidians in Coastal Water*,
DOI 10.1007/978-3-319-29118-5_3

Even though occurrence and distribution of ascidians are remarkable in various parts of the country, they are poorly studied from Indian coastal waters. It is, therefore, essential to explore this known species from unexplored areas of Indian coastal waters and economic importance of this group shall be helpful to draw a baseline for advanced research.

3.2 Diversity and Distribution

Information on species diversity, richness, evenness and dominance of the biological components of the ecosystem is essential to understand the healthy and wealthy status of the specific environment. Ascidian young ones, the tadpole larva, play a key role in dispersal and preservation of the species from being extinct. Dispersal and site selection of the larvae are believed to be the most important factors for the distribution of their races. The diversity of ascidians around the world is truly astonishing. The solitary ascidians *Ciona intestinalis* are found most temperate regions of the world. The members of the subfamily Botryllinae are widely distributed from the Arctic to southern Australia, New Zealand and South Africa. *Botrylloides perspicum* is widely distributed in tropical and temperate Pacific waters. Similarly, the solitary ascidian *Styela plicata* is a conspicuous component in marine communities in warm and temperate regions. A jet black simple ascidian, *Phallusia nigra*, native to Red sea is, widely distributed throughout the warm waters, tropical waters and temperate waters and confined to the sheltered waters, mostly inhabits harbour installations such as pilings, buoys, piers and also hulls of ships etc., In the harbour of Genoa, ascidians are abundant with *Ciona intestinalis,* a very important species in the quantitative composition of the sessile benthos.

Most of the tunicates live in the littoral zone, but few deep sea forms are known, such as *Hypobythius calycodes* found below 5000 m and *Cnemidocarpa bythius* from the hadal zone at a depth of 7000 m. The solitary ascidian, *Ciona intestinalis*, is one of the most abundant species with densities greater than 20,000 individuals/m^2 and *Microcosmus vulgaris* at a density of several hundred per m^2 on stones dredged from 120 m in the Gulf of Lions. The colonial ascidian, *Aplidium peruvianum* is one of the most common observed on the open coasts of Paracus, near Ballestas Islands and dominates benthic rocky communities at a depth of 5–10 m whereas the simple ascidian, *Herdmania momus*, is restricted and dominant in certain areas of Indian coastal waters and its distribution was recorded in between 5 and 25 m depth. Certain species of tunicates such as *Didemnum psammathodes, Eudistoma viride, E. lakshmiani, Lissoclinum fragile, Perophora multiclathrata, Diplosoma swamiensis, Phallusia nigra, Ascidia sydneiensis* and *Symplegma viride* settled throughout the year in Thoothukudi coast of India. The population of *Botrylloides* sp., and *Polyclinum* sp., shows marked seasonal variations. Generally, the senescence and recession period are noted during summer for the colonial ascidian *Polyclinum indicum*, whereas, late spring and early summer *for*

Botrylloides magnicoecum. The great majority of the species belonging to Styelidae and Didemnidae have continuous breeding activity.

The wide distribution of tunicates in a variety of habitats, substrates and varied depth indicates their multiple defensive strategies. One such strategy is reproduction, a kind of biological process by which ascidians maintain their progeny, which ensures the conservation of species. Tunicates show a great plasticity in reproduction limited sexual and unlimited methods of asexual reproductions. Even a small solitary animal of about 2.5 cm long can have functional gonads and full gonoducts. Generally, solitary species reproduce sexually whereas the compound ascidian reproduces sexually by internal fertilization and incubating the larvae in brood sac and asexually by budding and fragmentation. The fragmented colony has the ability to again reattach to the substrates facilitating the dispersal rates of species in a variety of habitats. Many ascidian species are capable of regenerating zooids from basal stolons and this occurs in *Ecteinascidia turbinata*. A solitary sessile tunicate, *Ciona intestinalis* spawns approximately 500 eggs per day and its larvae undergo metamorphosis in 1–5 days. The larvae of *Phallusia nigra* has to be settled within 3–12 h after fertilization. *Styela rustica*, a north polar species and *Pelonaia corrugata*-a boreo-arctic species breeds when the temperature is lower. The colonial ascidian, *Botrylloides leachi* breeds from May to July of every year. In Indian coastal water, *Eudistoma viride*, *Didemnum psammathodes*, *Lissoclinum fragile* and *Eudistoma lakshmiani* breeds throughout the year.

3.3 Factors Affecting Distribution

Recruitment of ascidians varies considerably both in space and season, which can have important consequences on population and community structure. While recruitment, certain factors affect the embryonic and post larval growth of ascidians. These factors can be broadly classified into two broad categories. They are intrinsic or biotic factors and extrinsic or environmental factors.

3.3.1 Biotic Factors

The major biological factors include intra and inter specific competitions for space, food and shelter. Predation pressure too affects the life history of ascidians (Eg.-*Botryllus schlosseri*). Even though tunicate employs several defensive mechanisms including the presence of calcareous spicules in the test and mantle bodies (*Lissoclinum fragile* and *Herdmania momus*) and the production of secondary metabolites (majority of colonial ascidians) and acidic tunic and high concentration of vanadium in the body, majority of the species are known to be affected by predators. Predation is a type of interspecific relationship where one partner is benefited and the other partner is harmed. Predators strongly impact communities of sessile invertebrates in intertidal rocky areas. Predation is the major cause for mortality of larvae and adult tunicates. So the larval period is a critical one when it enters into the new area where it has to be settled for its existence. The most important predators are crabs, polycheats worms, molluscs, certain aquatic mammals etc. A population of *Ecteinascidia turbinata* is controlled by polyclad worm *Pseudoceros crozieri*. The starfish, *Asterias rubens* feeds on *Ciona intestinalis*, *Pyura chilensis*, *Perophora multiclathrata* and *Didemnum psammathodes*. The prosobranch molluscs *Erata voluta* controls *Botryllus schlosseri and Botrylloides leachi*, and nudibranch molluscs *Goniodoris nodosus*, *G. custanea* and *Ancula cristata* feed on the colonial ascidians, *Diplosoma listerianum*, *Botryllus schlosseri and Botrylloides leachi*.

3.3.2   Abiotic Factors

Even though the effect of biotic factors drive to control the population of tunicates, abiotic factors in the habitat also influence the distribution and population of ascidians. The abiotic factors include temperature, salinity, light, dissolved oxygen, nitrogen content, turbidity, water current etc. Though there are several factors acting as limiting factors, substratum is a significant ecological factor which influences the distribution of organisms. The nature of the substratum is one of the factors determining the presence or absence of ascidian species. Selection of the available substratum is made mostly in the larval stages of the animals, so each distinct type of substratum will have its characteristic faunal and floral associations. A variety of substratum such as concrete submerged blocks, marine floats, pilings, buoys, pebbles, corals boulders, prolonged unused barges, oyster beds, ship's hull etc. provide suitable substratum for tunicates larvae to settle. The tadpole larval stage is more advanced than its adult, as the majority of the tadpoles favour smooth surface than rough surface. This indicates that selective ability and settlement nature of tadpoles is species specific. Habitat selection by settling larvae can be an important determinant of subsequent survivorship of adult's distribution and abundance. Ascidian tadpoles are able to differentiate the types of surfaces on which to settle, possibly due to the presence of mechanoreceptors found in the epidermal cells. For instance, several species of the Family Pyuridae live on rocks and some prefers soft bottom community. (Eg. *Ciona* sp. and *Ascidia mentula*—living on a soft muddy bottom). The sediment stability is an important habitat feature of *Molgula occidentalis* (Molgulidae). Certain species of Molgulidae, some in the Styelidae, and in the Pyuridae prefer loose sediments. Most of the solitary ascidians require hard substrates but few molgulids anchor mud or sand with hair like tunic protrusions.

There are several factors such as monsoonal flood, cyclonic storms etc. alter the nature of the substratum and topography of the substratum by depositing sand and silt, which prevent the attachment of sedentary forms in the intertidal zone, thereby alter animal population. The availability of different types of substratum enhances more settlement of ascidians. The colonial ascidian, *Didemnum psammathodes*, is recorded from all types of substratum. The majority of the simple ascidians, *Ascidia sydneiensis, Phallusia arabica, P. nigra, Microcosmus exasperatus* and *M. squamiger* usually prefer harbour area where they attach on hulls of ship, oyster beds and other permanent installations and found throughout the year; whereas, *Ascidia gemmata, Distaplia nathensis,* and *Eusynstyela tincta* colonize seasonally. So, the availability of substratum in the habitat impacts a positive significant effect on diversity of tunicates in macrobenthic community.

Next to substratum, temperature and salinity are the principal factors for ascidian as these two variables influencing recruitment and reproduction. Generally the *Ciona* sp. prefers the salinity range of 11–35 ‰. Only a few species can survive in salinities below 20 and 25 ‰ or above 44 ‰. *Ciona intestinalis*, a highly tolerant species survive in a wide range of salinities in between 12 and 40 ‰. Seasonal rain decreases the salinity of medium causing annual mortality of several species of the family Molgulidae. Antarctic species tolerate the temperature as low as −1.9 °C whereas others survive in seawater temperature above 35 °C in the Arabian Gulf. The temperature and salinity are not affecting the species *Didemnum psammathodes, Eudistoma viride, E. lakshmiani* and *Lissoclinum fragile* in Indian coastal waters. Didemnids are emerging worldwide as successful invaders and tolerate wide environmental stresses. The simple ascidian, *Styela plicata*, is a good model for the study of long distance dispersal of sessile species with short planktonic larval life-stages and has the ability to tolerate wide variation of salinity and temperature. If the new environment is conducive for their survival, particularly for feeding and breeding; they habituate well and spread rapidly in the new invaded habitats.

3.4 Invasiveness

Species suddenly moved into new environments often fail to survive, sometimes thrive and become invasive which in turn cause adverse effects to local flora and fauna through their toxigenic, proliferative and over competitive characteristics. As expressed in the Convention on Biological Diversity (2002), an exotic species could be considered as invasive when its propagation threats the ecosystems, habitats or other species, and consequently produces socio-cultural, economic, environmental or human health damages. Meanwhile, the recent investigations revealed that the marine environment has been encountering a number of problems such as dredging, overfishing, pollution, coral bleaching and changes in ocean current patterns and ever growing sea transportation cause the entry of exotic species from its native to unknown coastal areas resulting in homogenization of the global biota. The introduced species may compete with the native species for space, food, shelter etc., and gradually eliminates them by adopting the principle of survival of the fittest in the theory of natural selection. Among the several factors implicated for invasion of species, shipping is considered as the major anthropogenic vector for the entry of many species from egg to adult stage through ballast water or hull fouling. Ship's hull is very important potential source for the introduction of exotic sessile species and these species affect the quality and quantity of cultured organisms causing economical loss to the aquarists.

Besides, a gradual warming of climatic conditions due to global warming leads to successful establishment of entry of many exotic species into new habitat and facilitate quick acclimatization to changed environmental conditions. Ascidian's short-lived, non-feeding, pelagic tadpole larvae have played key role in dispersal of the species over long distances through ballast water. Over 80 % of the world cargo is mobilized transoceanically and over 12 billion tons of ballast water is filled at one part of the ocean and discharged at the other. This ballast water is conducive for a variety of macro- faunal larval stages and cysts to translocate into alien regions, usually along the coasts of the continents. As tunicates are fouling in nature, they can be transported to new areas by ships. Ascidian larvae are transported through ballast water while juveniles and adults are through hulls. Once the long—distance transport of a species has resulted in successful colonization of a new harbour, the species almost always seems to spread quickly to neighbouring harbours probably by fouling on hull of small crafts. Besides, they overgrow on organisms such as, hydroids, macro algae, sponges, bryozoans, mussels and oysters and even on other tunicates. As a result, the endemic species cannot compete with the exotic species mainly in respect of feeding, breeding and respiration. Generally ascidians are nuisance to mariculture operations by fouling the nets, culture cages, even on the oysters resulting in mortality of culturable species and loss to the mariculture personnels. The commonest widespread invasive species are *Didemnum* sp., *Styela clava* and *Botrylloides* sp. Many ascidians are highly invasive in nature and quickly spread to new habitats where they damage the coastal installations by their fouling nature and displace the local communities by its space-dominance nature. *Styela clava* is an aggressive invader to temperate marine ecosystems and invaded North America, Europe, Australia and New Zealand.

3.4.1 Control Measures

There are large numbers of specific methods such as mechanical, physical and biological method to control invasive ascidians. These methods suffer from one or other limitations and none of them is successful in completely removing the invaders from their environment since these are potential breeder. For instance, *Styela clava*, native to Northwest Pacific, *Didemnum perlucidum*, a cryptogenic species to Southern Brazil and *Eudistoma elongatum*, native to Australia are commonly widespread tunicates introduced into majority of the coastal areas, particularly they spread throughout the harbour areas and fouling on the aquaculture installations and culturable species. Although usually confined to harbours, in some cases these invaders can colonize natural subtidal areas and quickly overgrow and smother the native fauna, thereby significantly changing the benthic community. For instance, the large *Pyura praeputialis*, an invader to Chile from Australia, is now so abundant on rocky shores of Antofagasta Bay and completely modified the habitat, providing shelter for 116 species.

Due to dual mode of life (larva and adult) in its life history, it is very difficult to eradicate completely from the habitat. Therefore, hand picking and plucking methods are applied to remove adults from artificial structures under immersed condition. Besides, air drying, immersion in freshwater, encapsulation, acetic acid and chlorine exposure are applied for the fouling structures when on land. The larval mobility of *S. clava* and its settlement can be reduced by using chemicals such as medetomidine which is a potential tool for controlling the population. However, control or eradication of tunicates is often technically and financially difficult and time consuming. For example, acetic acid was effective to control *E. elongatum* but ineffective to *Didemnum* sp., Therefore, the entry of exotic species is inevitable in the years to come but maximum utilization of exotic species under natural conditions is the best way for promoting the economic status of one's nation.

3.5 Sentinel Ascidians

3.5.1 Bioindicator

Marine ecosystem is most probably prone to stress condition as it receives pollutants through industrial effluents, agricultural wastes and land run-off waters. This polluted ecosystem greatly affects the marine organisms particularly the sessile forms. Many sessile organisms have developed various mechanisms to get rid off these stresses and survive better. Benthic organisms are frequently used as bioindicators for trace contaminants as they are closely contact with the sediments. Interestingly to note that ascidians are one such group that has the ability to accumulate overindulgence of metals in their body tissues than that of environment. Being benthic, sedentary and filter feeder, ascidians receive pollutants along with food particles by the process such as adsorption, precipitation and accumulation particularly in the blood due to the presence of specilized cells, vanodocytes and the pigments tunichromes. Absence of kidney also ensures the accumulation of metals in the body. For that reasons, ascidians have been recognized as a potential candidate species for bioaccumulation of metals. These attributes help the environmentalist to determine the level of pollutants in water as well as in its body. Ascidians could be recognized as amazing creatures which are used as excellent biological markers for metals in aquatic environment. With the importance of these superorganisms in the marine environment, they are classified into three broad categories based on the adoptability to environmental stress. They are regressive species, transgressive and tolerant species. Regressive species are also known as sensitive species because of their increased sudden disappearance with increasing the environmental stress in the habitat. Eg. *Herdmania momus*. They are ecologically specilized. Transgressive species are dominant in sheltered areas such as harbour and nearby areas where water replenishment is low. They are dominant biofoulers in their habitats (Eg. *Microcosmus squamiger)*. Tolerant species are capable of living even in high stressed areas but decrease in populations in the most anthropogenic locations cannot be ruled out (Eg. *Botryllus schlosseri*).

3.5.2   Metal Accumulator

Due to great deal on presence of specialized tunichrome and vanodocytes, in the blood of tunicates, they are grouped into iron and vanadium accumulators. Most species of ascidians are capable of accumulating the metals in small amounts but ascidians belonging to the families of Ascidiidae and Perophoridae of the sub-order Phelobobranchia are found to accumulate the metals in large quantity. Notably, some species of ascidians (*Phallusia nigra and Ascidia sydneiensis*) show a spectacular concentration of vanadium with a range as high as 3700 ppm dry weight in ascidians against the range of 0.3–3.0 µg in seawater. Extraordinarily, some ascidians accumulate vanadium in excess of one million times as high as that in sea water. Whereas in the family Pyuridae of Phleobobranchia, iron appears to be concentrated more than vanadium, niobium, tantalum and titanium (Eg. *Eudistoma ritteri*). The phlebobranches principally store the metals in all their tissues and are the best indicators for metallic and organo metallic pollution. Even the specific parts of body also have the ability for selective absorption of various metals such as titanium, vanadium, chromium, manganese, iron and copper (e.g. The ovary of *Ascidia malaca* and *Ciona intestinalis* and in gonads of *Polycarpa gracilis*). Some members of the families Cionidae and Diazonidae have vanadium in their tissues but not in blood. The black ascidian *Phallusia nigra* accumulates cadmium, mercury and lead in both the test and mantle bodies. In the port area of Thoothukudi coast in India, the tunicates such as *Styela canopus, Microcosmus exasperatus* and *Polyclinum constellatum and Phallusia nigra* accumulates cadmium, copper, nickel, lead, zinc, manganese, cobalt and iron but showed variations in the accumulation. Likewise, some ascidians can be used as an effective bioindicator of metals such as Cd, Pb, Hg and V. Recent investigation revealed that the sudden disappearance of *H. momus* from harbour area of Thoothukudi coast, India could probably be due to the higher concentration of metals in this area due to ever handling of more thermal coal, fertilizers, raw materials like rock, phosphate, timber and copper. Translocation of this species towards nearby areas indicates the prevalence of variable stressed conditions which might have disturbed this transgressive species.

Chapter 4
Association

4.1 Associated Fauna

One of the fascinating activities of the ascidian population has received much attention towards the ascidiologists is co-adapting nature of ascidians with other organisms; even though the physical and chemical defences exist. Ascidians have been given the status of "superorganisms" as they provide space, food and shelter to support inter and intra-specific relationship. The life history traits of tunicates force them to cohabit with other animals gradually by means of natural disturbance and the effect on space allocation, recruitment, competition and subsequently direct interspecific relationship. Therefore, ascidians are considered as keystone species in the benthic megafaunal community. Their loss from a conservation area leads to the loss of numerous other species as inadvertently reserved other species. The change in ascidian composition echoes the changes in environmental variables as well as community composition.

Predation, parasitism and commensalisms are important aspects of ascidian biology and ecology with the intention that ascidians play an important role for the evaluation of different behavior of the associated species. There are some species whose occupancy area is large enough and habitat requirements are wide enough particularly for food, survival and shelter. They also bring other species intentionally or coincidently under protection. The simple ascidian, *Herdmania momus* shows commensalisms like partnership with the co-partner. Even though the simple ascidian *H. momus* has predominantly barbed spines in the test, its larger body size with a rough wrinkled test and enlarged folded branchial sac provided tranquil home for many epizoic (barnacles and encrusting tunicates) and endolithic (*Modiolus* sp.,) organisms and acted as umbrella species. Hence, *H. momus* can be

© Springer International Publishing Switzerland 2016
H.A. Jaffar Ali and M. Tamilselvi, *Ascidians in Coastal Water*,
DOI 10.1007/978-3-319-29118-5_4

used as an ideal flagship species to study faunal association, environmental assessment and in the preparation of environmental sensitivity maps to monitor the marine environment. Similarly the amphipod, *Leucothoe commensalis* has commensalistic association in peribranchial cavity of *Phallusia nigra* (Fig. 4.1).

All kinds of associations are exhibited by ascidians with diverse group of animals in different parts of the world. Several common associated organisms are polychaetes in the cloacal cavity of *Microcosmus multitentaculatus,* zoochlorellae in the lacunae of the common test of *Trididemnum cyclops*, nemertine *Tetrastemma cillalum* in 90 % of *Microcosmus sabatieri*, amphipod *Leucothoe commmensalis* in peribranchial cavity of *Phallusia nigra* and *Herdmania momus*, *Leucothoides pottsi* in the tunicate *Ecteinascidia turbinata* and the bivalve, *Musculus mamoratus* in the test of *Ascidia mentula*. Several individuals of Dorophygus, amphipods, polychaetes and even pinnotherid crabs are found to live inside the branchial and atrial cavities of many ascidians.

Herdmania momus is encrusted with over 40 individuals of mussels. The bivalve *Mytilimeria nuttalli* Conrad, 1837 is embedded in a variety of colonial ascidian species. Several individuals of *Notodelphys allmani* often inhabit even small specimens of *Ascidiella aspersa*.

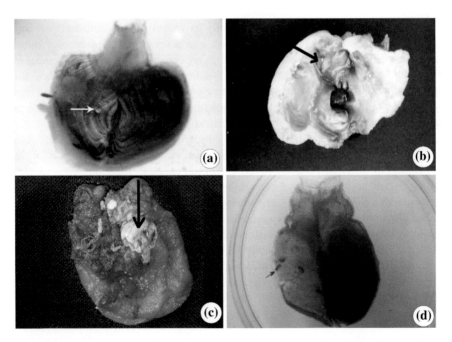

Fig. 4.1 Associated fauna in ascidians. **a** *Leucothoe commensalis* in peribranchial cavity of *Herdmania momus.* **b** *Modiolus metcalfei* in hypodermic test of *Herdmania momus.* **c** Epilithic of barnacles on the test of *Herdmania momus.* **d** *Leucothoe commensalis* inside the peribranchial cavity of *Phallusia nigra*

4.2   Associated Flora

Ascidians are exclusively marine, sedentary and filter feeding animals and consume large amount of planktonic organisms from the medium. The water enters through the branchial siphon to the atrial siphon via enlarged perforated pharynx with gills brings large quantity of suspended matter and bacteria.

The gut content of ascidians contains inorganic materials, small brown cells and many bacteria. The incidence of bacterial flora in the alimentary tract not only depends on the filter feeding habit but also on food and digestive pattern of the ascidians. The solitary ascidians are classically believed to feed on phytoplankton, detritus and bacteria. The gastro intestinal bacteria may play a vital role in nutrition, digestion, growth and disease susceptibility of ascidians. The heterotrophic bacterial flora of ascidians is found to be influenced by the flora present in the environment.

Prochloron (phylum cyanobacteria) is a unicellular photosynthetic prokaryote organism commonly found as an extracellular symbiont on coral reefs and in didemnid ascidians (sea squirts). Since Prochloron is an obligate symbiont, it requires a symbiotic relationship to thrive and survive. The ascidian-prochloron symbiosis is undoubtedly a mutual symbiosis. The cloacal cavity of the didemnid colony, habitat for Prochloron, is a highly protected space against predators. In ascidian-prochloron symbiosis, the benefit of the host ascidians has been thought to be photosynthetic metabolites provided from the symbionts. There are some reports of radio-labeling studies showing that Prochloron cells transfer photosynthetically fixed carbon to the host as soluble compounds of low molecular weight (Pardy and Lewin 1981; Griffiths and Thinh 1983). These symbionts, including *Prochloron* sp., may control the fouling and colonization with some compounds, and this would be beneficial to the host ascidians that suffer uncontrolled fouling. Transfer of nutrients is the main reason for the symbiotic relationship between prochloron and didemnids. There is also some relevant data that shows there is nitrogen exchanging between the host *Leptoclinides patella* and the symbiont *Prochloron didemni*, as well as nitrogen fixation. In tropical and subtropical waters, some colonial ascidians belonging to the Family Didemnidae have obligate symbiotic relationships with Prochloron, Synechocystis, and/or unknown cyanophytes. Kott (2001) described about 30 didemnids as host species. A larger number of photosymbiotic ascidian species are generally distributed at lower latitudes, and thus the distribution of these species is a potential indicator of the warming of seawater. Thus, water temperature may be crucial in limiting the distribution of photosymbiotic ascidians. On the other hand, an increase in water temperature caused by global warming may alter the range of species in the near future.

Chapter 5
Commercial Importance

From the last two decades, scientists have paid great attention on ascidians, the phylogenetically important species belong to the subphylum Urochordata, as they contribute more in the field of biotechnology, biomedicine, aquaculture, biochemistry, microbiology, immunology etc. It is well understood that every species makes some contribution to the structure and function of its ecosystem. In recent decades, marine habitats have come under increasing stress due to various activities such as habitat destruction, land based pollution, shipping, coral mining, dredging etc., which in turn cause upset the number and biomass of fauna and flora, sequentially affects the food chain and causes serious threats to the sedentary organisms particularly on ascidians. As the ecological and economic value of habitat are mainly dependent on the functions of the species, the diversity, distribution, its sustainable utilization and conservation of this species has to be focused in future for the benefit of the mankind.

5.1 Food Source

In contrast to the fear of ever increasing invasive species in marine ecosystem all over the world, not all invasive species are harmful as some members of this group have been identified as non-conventional food resource from the marine ecosystem. In the natural environment, tunicates are prey for a number of animals including flat worms, annelids, arthropods, molluscans, echinoderms, sea birds, sea otters etc., and some ascidians are used as food for human consumption in various parts of the world such as Chile, France, Korea, Italy, Japan etc., Like crustaceans, molluscans and finfishes, ascidians are also used as food and supplementary feed for culturable fishes in aquaculture. They have been considered as a possible source of cellulose,

© Springer International Publishing Switzerland 2016
H.A. Jaffar Ali and M. Tamilselvi, *Ascidians in Coastal Water*,
DOI 10.1007/978-3-319-29118-5_5

protein, aminoacids and minerals. Besides, ascidians have appreciable amounts of biologically important minerals such as sodium, potassium, phosphorus, iron, magnesium and manganese. The finest quality of food such as colour, texture, flavour, odour etc., depends on the amount of aminoacids present which in turn, reflects the high quality protein in the food. The essential aminoacids (Valine, Threonine, Methionine, Isoleucine, Leucine, Histidine, Lysine, Tryptophan, Arginine and Phenyl alanine) and non-essential aminoacids (hydroxyproline, Asparagine, aspartic acid, glutamic acid, alanine, glycine, serine, cystine, tyrosine) are found in many solitary ascidians.

Majority of the species belongs to the family Pyuridae are consumed as food. There are certain species that are cultured and harvested in various parts of the world. They are eaten as raw or as ingredients, or in the processed forms when their productivity is surplus amount. *Halocynthia roretzi* is an edible ascidian and eaten as raw after removing the internal organs and slicing the animal vertically, serving them with vinegared soy sauce. In Greece, they are consumed as raw food with lemon or in salads with olive oil, lemon and parsley. In Korean's restaurant, the *Styela clava* is a traditional special food item, prepared by steaming Styela with hot pepper paste and garlic with some vegetable condiments. For trading purpose, the surplus farmed tunicates are salted, smoked, grilled, deep-fried or dried to extend the shelf life of the products. This tunicate is farmed successfully in Japan and Korea. The simple ascidians, *Microcosmus sabatieri* and *M. vulgaris* in Europe, *M. hartmeyeri* in Japan, *M. sulcatus, Styela plicata* and *Polycarpa pomaria* in Mediterranean and *Pyura chilensis* in South America, *Microcosmus sabatieri, Styela plicata* and *Polycarpa pomaria* and several similar species from the Mediterranean Sea (Harent 1951) are the popular food items. In Chile, *Pyura chilensis* is an edible species and overexploited for consumption as it contains more amount of iron. *Styela clava* is a delicate food in Southern Korea and exported to U.S countries in frozen form and *S. plicata* is used in fish paste.

Several species of tunicates are commercially important to aquaculture industry. Sea pineapple (*Halocynthia roretzi*), *H. aurantum* and *Microcosmus hartmeyeri* are cultivated and eaten locally in Japan (hoya, maboya) and Korea. *Pyura chilensis* is heavily fished in Chilean coast and exported this species to Sweden (32.5 %) and Japan (24.2) countries to earn foreign exchange. In Australia, *Pyura praeputialis* is used as bait for fishing where *Pyura stolonifera,* known as cunjevoi, was used as food source by a group of local people and now this species is used as fishing bait. The supplementary feeds prepared from the simple ascidian, *Herdmania momus* and the colonial ascidian, *Didemnum psammathodes* increased the yield of black-molly *Poecilia sphenops* and also for other culturable species in India. Ascidian pickle was prepared from *H. momus* and canned for six months for storage successfully (Fig. 5.1).

Fig. 5.1 Food source **a** pellets prepared from *Polyclinum indicum* **b** pellets prepared from *Herdmania momus* **c** huge quantity of *Herdmania momus* **d** pickle prepared from *Herdmania momus*

5.2   Pharmacological Source

Most of the benthic organisms produce potential bioactive metabolites in response to various ecological stresses such as competition for space and food, maintenance of unfouled surface and deterrence for predation to live successfully. Likewise, ascidians are also known to contain a variety of novel and highly potent bioactive compounds, which have been hypothesized to function as chemical defence. As chemical defences, many sea squirts intake and maintain extremely high concentration of vanadium in the blood, acidic pH in easily-ruptured bladder cells in the tunic, and (or) produce secondary metabolites to avoid predators and invaders. Some of these metabolites are toxic to cells and are of potential use in pharmaceuticals.

In pharmaceutical industry, the drug development process includes screening, identification, efficacy testing, safety testing and large scale of commercial production for marketing. Formerly, scientists targeted the micro-organisms, plants and animals from the terrestrial ecosystem and its yield was the only 1 out of 10,000–20,000 molecules reached the market, which may take 10–15 years and cost up to $800 million in markets. This has resulted in large pharmaceutical groups abandoning their search for new drugs derived from natural substances. The organisms particularly the sessile forms with the high degree of "survival of the fittest" can produce the large quantity of secondary metabolites and hence the research community has taken the efforts to screen the biologically potent compound from the sessile organisms such as sponges and bryozoans. Later, they have recognized the importance of ascidians in the same field as they lead a successful life in stressed areas and accountable for producing secondary metabolites towards chemical defence.

So far, over 20,000 bioactive compounds have been derived from various marine animals including tunicates. Large number of natural products has been isolated from ascidians and tested for various biological activities such as antimicrobial, antineoplastic, antitumour, antifouling, antioxidant, anti-inflammatory, etc. In addition, various nitrogenous and non-nitrogenous metabolites have been isolated and tested for plant growth regulatory activity, deterrent activity, insect control, wound healing activity, hepato productive activity, immune stimulating activity etc.

With regards to natural products extraction, members of Aplousobranchia, particularly Clavelinids, Polyclinids, Polycitorids and Didemnids are the best represented group, with more than 80 % of ascidian species from which natural products have been isolated. It is noteworthy to understand that 90 % of the reported secondary metabolites from tunicates are nitrogenous with many compounds displaying remarkable bioactivity. More than 300 alkaloid compounds have been isolated from marine ascidians and they have pharmacological activities. Majority of the alkaloids have been isolated from the member of the genus *Eudistoma*, followed by *Ritterella*, *Pseudodistoma*, *Didemnum*, *Synoicum* and *Lissoclinum*. Tyrosine is precursor for a large number of alkaloids. Ascidians have also been the potential source of proto alkaloids, terpenoid and steroids. More than 20 % of natural products derived from ascidians are non-nitrogenous compounds. These compounds are mostly terpenoids or steroids.

5.3 Drugs of Ascidians in Clinical Trials

Before introduce a drug into a commercial market, it is significant to have pre-clinical and clinical trials to understand the pharmacology of new compounds, dose effect and other effects on cells or organisms. Nature provides a lot of potential treasure of medicinal value based organisms, one such organisms is tunicates.

The biologically potent Didemnin B was originally isolated from the Caribbean tunicate *Trididemnum solidum* and was the first marine compound to enter human cancer clinical trial as a purified natural product. Mechanistically, Didemnin B interrupts protein synthesis in target cells by binding non-competitively to palmitoyl protein thioesterase. Aplidine isolated from the Mediterranean tunicate *Aplidium albicans* exhibited anticancer properties in preclinical animal tests. The marine natural product Diazonamide A, extracted from the Philippine ascidian *Diazona angulata*, possess potent microtubulin interactive activity arresting the process of cell division in cultures exposed to treatment.

Ecteinascidin 743, a tetrahydroisoquinoline alkaloid, isolated from the Caribbean Sea squirt *Ecteinascidia turbinata*. ET-743 was active against a range of tumour types particularly against solid tumours in standard animal models. Subsequent human trials showed efficacy against advanced soft tissue sarcoma, renal carcinomas, osteosarcoma and metastatic breast cancers etc. Vitilevuamide, one of the several novel tubulin interactive agents, is a bioactive cyclic peptide isolated from the ascidians *Didemnum cuculiferum* and *Polysyncraton lithostrotum*. Vitilevuamide inhibits tubulin polymerization and can arrest the cell cycle of target cells in the G2/M phase.

Varacin (4) is another well-known marine polysulfide natural product, which contains a pentathiepin ring and notably has a similar carbon framework to dopamine. Varacin has shown potent antifungal activity as well as cytotoxicity towards human colon cancer 100 times that of 5-fluorouracil that was isolated from the ascidian *Lissoclinum vareau* collected from the Fiji islands.

Chapter 6
Biopotentiality of Compounds of Ascidians

6.1 Antibacterial Activity

By virtue of their sedentary and filter feeding habits, ascidians accumulate high concentration of bacteria in branchial siphon. In order to get rid of microcosm, ascidians synthesis various kinds of antimicrobial compounds. It has been proposed that antimicrobial substances function as humoral factors in defence mechanisms of invertebrates, which lack humoral immunoglobin. Inhibition of bacterial growth has been one of the methods used to test the degree of toxicity of bioactive metabolites. A well-known microbial antibiotic, enterocin, against *Saccharomyces cerevisiae* and *Candida albicans* was isolated from *Didemnum* sp. Polycarpamine B was isolated from the solitary ascidian, *Polycarpa auzata*. A powerful antibiotic, octapeptide, was isolated from the tunicate, *Styela plicata* against *Staphylococcus aureus*. Styelin D inhibits the growth of both Gram-positive and Gram-negative bacteria and exhibited hemolytic and cytotoxic properties against eukaryotic cells.

6.2 Antitumour Activity

The ability of chemical to kill rapidly dividing cells is the hallmark of chemotherapy. The first group of antitumour marine compound, botryllamides was isolated from the ascidians, *Botryllus* sp. from Philippines and *Botryllus schlosseri* from the Great Barrier Reef of Australia. They showed mild cytotoxicity against the human colon cancer cell line HCT-116. Later, the ecteinascidins- a potent bioactive compound isolated from the colonial ascidian, *Ecteinascidia turbinata*, has strong

© Springer International Publishing Switzerland 2016
H.A. Jaffar Ali and M. Tamilselvi, *Ascidians in Coastal Water*,
DOI 10.1007/978-3-319-29118-5_6

antitumour properties. Majority of the members from family Didemnidae contains different potent bioactive compounds (Didemnin B) effective against fibroblasts and tumour cell lines. A cytotoxic compound Bistratene A is a polyether compound obtained from *Lissoclinum bistratum*.

Although a number of screening efforts have indicated a much higher percentage of antineoplastic/antitumour activity in marine organisms than in terrestrial plants, only recently have marine natural products made their first appearance in clinical trials at the National Cancer Institute, Maryland—first, the didemnins, and then the bryostatins from ascidians (John 1986). One promising group of bioactive metabolites is ecteinascidins, biosynthesized by the colonial ascidian, *Ecteinascidia turbinata*, which has strong antitumour properties.

6.3 Antifouling Activity

As fouling organisms, ascidians contribute significantly to cause global issue of marine biofouling on the hulls of ships, wood and metal pilings barrels, nets, hard rocks, pebbles, cobbles, boulders, cement blocks, barges etc. They over grow on organisms such as sponges, macro algae, hydroides, anemones, bryozoans, mussels and oysters, and even other tunicates. Since ascidians are intense competitor for space on intertidal rocky substrates, the presence of growth regulators in this group certainly seems plausible. The tunicate extract (BT-ll) was a potent inhibitor of root growth in seedlings. As per the available literature, among the 52 members of genera of Aplousobranchia, about 20 members showed potential antifoulant activity, followed by Stolidobranchia (18 %) and Phlebobranchia (12 %).

6.4 Insecticidal Activity

Antimalarial compounds have also been isolated from the solitary ascidians, *Microcosmus goanus*, *Ascidia sydneiensis* and *Phallusia nigra*. The tunicates such as *Clavelina picta*, *Eudistoma obscuratum*, *Didemnum molle*, *Atriolum robustum*, *Phallusia* sp., *Didemnum* sp. and *Aplidium* sp. showed insecticidal activity against *Anopheles maculatus* and *Aedes aegypti* indicating tunicates are as effective as the synthetic insecticide in killing the mosquitoes.

Chapter 7
Ascidians—Model Organisms

Ascidians represent an interesting model organism to study in various points of view. For many years, there has been a subject on discussion of taxonomic position of Urochordates. There are certain characters, which resemble both invertebrate and chordate. For example, the larval stage possess a notochord in the tail region which resembles the chordate characters, whereas the adult stage do not contains notochord, which is resemble the invertebrates. Hence, the members of this group are kept in a separate subphylum Urochordata. Besides, sedentary mode of life, filter feeding, hermaphrodite gonads and asexual reproduction strongly support to keep ascidians in the phylum invertebrata. In contrary, the presence of notochord in the larva, gill slits in the pharynx and heart with pericardium support to place this group in the Phylum chordata. In addition, ascidian has some specialized characters of their own viz: sedentary mode of life in adult, non-living test and retrogressive metamorphosis in the life history. So, in the evolutionary point of view, it is an important group that connects invertebrata and chordata. Hence, they are said to be a connecting link between invertebrates and chordates. An interesting fact that asexual mode of reproduction such as budding and fragmentation can be seen in lower group (invertebrates) of organisms whereas the sexual reproduction is a common rule in higher organisms. In tunicates, a spectacular reproduction i.e., alternation of generation is also adopted. The colonial ascidian, *Botryllus schlosseri*, represents a model organism to study the mode of reproduction (sexual and asexual) and colonial life style.

There are some members of the ascidians which can be used to study as model organisms for biological invasion. (e.g. *Didemnum vexillum, Ciona intestinalis* and *Styela clava*). These three species can tolerate to live in changed environmental variables and reproduce their young ones at a faster rate. For instance, *Styela plicata* is a good model for the study of long distance dispersal of sessile species with short planktonic larval life-stages. Even the small size of the mature adult reproduces about 5000 eggs which hatch after 12–15 h. The non-native tunicates *Didemnum vexillum*, threaten the diversity of native species of *USA and* shellfish aquaculture in other regions.

© Springer International Publishing Switzerland 2016
H.A. Jaffar Ali and M. Tamilselvi, *Ascidians in Coastal Water*,
DOI 10.1007/978-3-319-29118-5_7

Regeneration is the ability to form an entirely new individual from the peripheral vasculature. Having been a close phylogenetic relationship to the vertebrates, the ascidians are known for both embryonic and regenerative studies. Colonial ascidians are the only chordates to undergo whole body regeneration. Colonial ascidians belong to the genus *Botryllus*, *Botrylloides*, *Perophora*, and *Clavellina* are able to regenerate from small fragments or pluripotent blood cells. *Botrylloides violaceus*, a colonial ascidian, regenerates after ablation of all zooids and buds of young colonies. In solitary forms, ablated tissues are replaced. For example *Ciona intestinalis* has potential to completely regenerate neural complex, oral siphons, neurons and multiple tissues after surgical removal. In few members of the genus *Polycarpa*, the internal organs such as peribranchial lining and connective tissues (tunic) can be reconstituted. The rapid development of both solitary and colonial ascidians offers amenable model system to study the molecular mechanisms which underlie cell fate decisions.

Part II
Ascidians of Southern Indian Waters

Chapter 8
Introduction to Indian Ascidians

India is one among the 12 mega biodiverse countries on the globe. It is a matter of great pride that its unique geographical location, varied climatic conditions and diverse habitats both natural and artificial in different latitudes promote the settlement of ascidians. Besides, the peninsular India, encircled with Arabian Sea, Indian Ocean and Bay of Bengal, forms a part of the principle sea route for many countries. Further Indian coast is being dotted with 12 major and number of minor ports, one might expect the entry of high rate of ascidians either as adult or larval forms through physical agents such as ballast water, hulls of ship, mechanized boats etc. that contribute positive significant level of diversity in this typical of tropical regions.

Recently, Gulf of Mannar in Indian coastal region has been recognized as Marine Biosphere Reserves on World Network by the UNESCO for its greater megadiversity region in the world. It is one of the world's richest regions from marine bio diversity perspective and the first marine Biosphere Reserve in Southeast Asia. This region, enclosed by the Bay of Bengal, is known to harbour diverse flora and fauna, making it one of the richest biodiverse coastal regions in Asia. But, it is pitiable to note that only 30 % of marine fauna have been identified in Indian coasts and the remaining is unexplored due to the insufficient collection at various depths, lack of professional taxonomists and unawareness of marine resource.

Taxonomy, the basic science of systematizing living organisms has been neglected in India for past few decades. However, of recent, its importance is being felt, particularly due to assessment of biodiversity. Unfortunately, only few experts in the taxonomy of marine organisms are available today in this country. Lack of job opportunities and the strenuous exercise involved in collection and preservation of the material, and lack of necessary funding may be the major reasons in this regard. Early knowledge on the flora and fauna needs updating to have a better understanding of the present status of the biodiversity. To attain this goal, more intensive collections of materials as well as literature reviews are needed. In this scenario, ascidians are gaining significant role in raising the biodiversity of Indian marine ecoysystem.

In India, diversity of ascidians has been studied over 2000 km off the coast line from Bombay to Vishakapattanam encompassing the regions such the Arabian Sea, Indian Ocean, Palk Bay, Palk Strait, Gulf of Mannar and Bay of Bengal by various authors (Oka 1915; Das 1938, 1945, 1957; Sebastian 1952, 1954, 1955, 1956; Renganathan 1981, 1982a, b, 1983a, b, c, d, e, 1984a, b, c; Meenakshi 1997, 2004;

© Springer International Publishing Switzerland 2016
H.A. Jaffar Ali and M. Tamilselvi, *Ascidians in Coastal Water*,
DOI 10.1007/978-3-319-29118-5_8

Meenakshi and Senthamarai 2004, 2006, 2013; Abdul Jaffar Ali and Sivakumar 2007; Abdul Jaffar Ali et al. 2009, 2010; Tamilselvi 2008).

Comparison of the pre-1990 ascidian survey data in Indian waters with that of the post-1990 period showed that more than 350 species of ascidians including 100 new species were reported in the post-1990 period and the total number of ascidians species increased to more than 400 though more than 3000 species of ascidians are documented in various parts of the world. Hence, a thorough knowledge is essential on this group on identification, classification, diversity and distribution which are central dogma for promoting research on pharmacogenomics, aquaculture etc.

Considering ongoing changes in marine ecosystem, the arrival and proliferation of non-indigenous ascidians is expected in future. In light of the above, it is imperative to investigate the ascidian fauna along the coastline of India and to create up to date check list of the region.

The present account provides the systematic review of the ascidians, main regions of highest diversity of ascidians and summary of the current trend of ascidian distribution pattern in India. We believe that this dataset is a significant contribution to the knowledge of management system of the ascidian fauna of this region and will be useful for future studies on ascidians and their habitats.

Chapter 9
Inventory of Ascidians of Southern India

9.1 Spatial Coverage

A total of 11 stations located along southwest and southeast coastline (Figs. 9.1, 9.2 and 9.3) were surveyed in natural and artificial substrates (Table 9.1). Samplings were made at various marine habitats such as intertidal flats, rocky intertidal zones, shallow water, deep sea, sandy beach, muddy flat, harbour/port water, shellfish reefs etc.

9.2 Sampling Description

Intertidal flats were visited during low tide level by walking access to collect samples available. Description of each substrate is briefed below:

Molluscan shells: The samplings were made from broken and dead molluscan shells present in the shallow water region and intertidal flat of Kayalpattinam station (Fig. 9.4a, b).

Oyster cages: Oyster cages are installed at a depth of 4–5 m by the Central Marine Fisheries Research Institute (CMFRI) at Thoothukudi Harbour. The collections were made from these cages during brushing (Fig. 9.4c).

Coral pieces: Many dead coral pieces are found to occur in the shallow water and intertidal flats of Mandapam and Hare Island stations. They were also sampled (Fig. 9.4d).

Embedded rocks: Small to large boulders are embedded partly in shallow water regions and intertidal flats. The samplings were made at the submerged portions of the boulders by snorkeling (Fig. 9.4e).

Calcareous stones: The calcareous stones of varying sizes ranging from 1–5 m^3 are laid down to break the sea water. These stones are considered here as artificial substrates (Fig. 9.4f).

© Springer International Publishing Switzerland 2016
H.A. Jaffar Ali and M. Tamilselvi, *Ascidians in Coastal Water*,
DOI 10.1007/978-3-319-29118-5_9

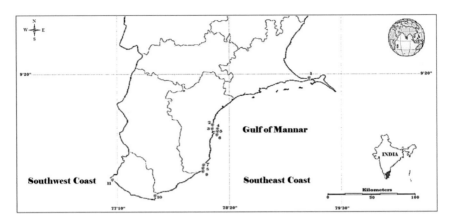

Fig. 9.1 Map showing the stations where ascidians sampled

Fig. 9.2 Survey areas **a** Mandapam **b** Muthunagar Beach **c** Inigo Nagar **d** Hare Island **e** North Break Water

Fig. 9.3 Survey areas **a** Thoothukudi Harbour **b** Kayalpattinam **c** Veerapandianpattinam **d** Tiruchendur **e** Chinna Muttom **f** Colachel

Harbour Installations: Cement blocks of varying sizes, pillings and ropes were also sampled for ascidians (Fig. 9.4g, h).

Mussel bed: The stations Thoothukudi Harbour and Veerapandianpattinam are known for molluscan fishery. Molluscan beds of varying size are available at a depth of 10–15 m. Generally brown mussel, green mussel and some other bivalves are seasonally fished at these stations. Observations were made on the shells of these molluscans by SCUBA diving.

Hull of boat: In the harbour stations the observations were made on the hull of fishing boat and barges by snorkeling and SCUBA diving.

All substrates (natural and artificial substrates) at the surveyed sites were equally searched and when appropriate, small boulders were overturned and examined. Since many ascidians have been trapped unintentionally during fishing at 15–20 m

Table 9.1 List of sampling stations along the Southern Indian coastline

Stations	Coordinates	Substrates
Mandapam	9°17′29″N 79°9′1.1″E	Hull of boat, harbour installations, rope, coral piece
Muthunagar beach	8°48′38″N 78°9′53″E	Calcareous stones
Inigo nagar	8°47′22″N 78°9′37″E	Calcareous stones
Hare island	8°46′36″N 78°11′38″E	Calcareous stones, coral piece, Embedded rocks
North break water	8°46′24.3″N 78°12′5.6″E	Calcareous stones
Thoothukudi harbour	8°45′14″N 78°12′39″E	Calcareous stones, hull of boat, oyster cage, cement block, mussel bed
Kayalpattinam	8°34′15″N 78°7′15″E	Molluscan shell
Veerapandianpattinam	8°32′08.6″N 78°07′17.0″E	Mussel bed
Tiruchendur	8°29′49″N 78°7′32″E	Embedded rocks
Chinna muttom	8°5′47″N 77°33′50″E	Calcareous stones, embedded rocks, cement block
Colachel	8°10′19″N 77°15′14″E	Hull of boat, cement block

depth, fish landing centers were also visited to examine *D. psammathodes* from the fishing nets and trawl.

At shallow water regions, small boulders were overturned, examined and then attached ascidians were sampled. Snorkelers were engaged to collect ascidians from large embedded rocks, boulders, harbour installations, jetty etc. at a depth of 2–4 m. Since many deep sea ascidians have been trapped unintentionally during deep sea fishing, fish landing centres were also visited to collect ascidians from the fishing nets and trawl. Both simple and colonial ascidians were collected from the undersides of floating docks and barges by engaging SCUBA divers at marinas. All possible collecting methods such as hand picking, scraping, peeling-off, dislodging the whole animal etc. were employed.

9.3 Identification

With experience, some species of ascidian can be identified in situ and others were collected, photographed and transferred in a container filled with sea water. The collected specimens were narcotized and then preserved in 8 % formalin with seawater. In case of sampling of some large colonial ascidians, a portion of it was

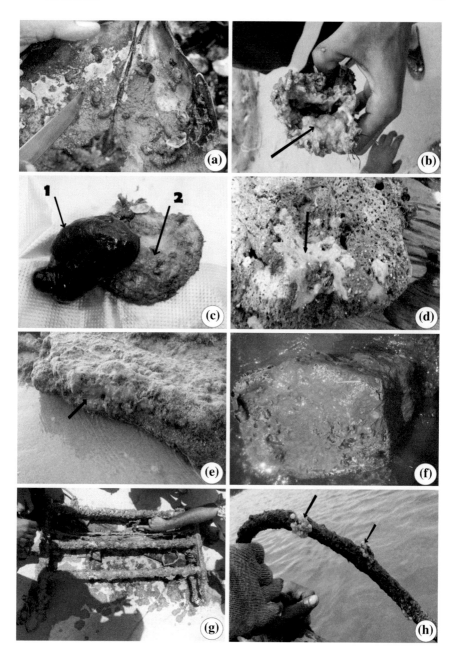

Fig. 9.4 Substrata **a** Bivalve **b** Molluscan shell **c** Oyster **d** Coral piece **e** Partly embedded rock **f** Calcareous stone **g** Harbour installation **h** Rope

collected after noting the structure and dimension of the whole colony. In case of synacidians, its representative was collected. The specimens were sorted and identified to species or the lowest practicable taxon with dissection, compound and stereo microscopes using various taxonomic references (Tokioka 1967; Miller 1975; Kott 1985, 2001; Monniot and Monniot 1996). Voucher specimens were housed at Zoology Museum of Islamiah College (Autonomous), Vaniyambadi. In the present study, the following both morphological and anatomical features of solitary, colonial and synascidians were thoroughly examined and analysed for identification of ascidians.

9.3.1   Taxonomical Structures of Solitary Ascidians

Appearance of test	wrinkled, smooth, furrowed, creased, colour, thickness, translucence, inclusions, outgrowths and other decorations, epibionts, blood vessels
Apertures	form, position and its siphonal armature
Branchical tentacles	simple or branched
Body muscles	arrangement, position, number, inner and outer layers
Dorsal tubercle	position, size, form of slit
Dorsal lamina	plain membrane, languets
Branchial sac	flat or folded, number of folds, stigmata per mesh, shape of meshes, shape of stigmata, number of internal longitudinal vessels, parastigmatic vessels and papillae
Gut loop	position, length and form of subdivisions, stomach folds, caecae, liver diverticulae, attachment to parietal body wall, presence of endocarps, and condition of anus
Gonads	number, arrangement, shape, position and attachment to body wall, position, orientation and length of gonoducts and relationship of ovary to testis

9.3.2   Taxonomical Structures of Colonial Ascidians

Test	thickness, consistency, translucence, colour, inclusions, epibionts, surface sculpture and/or decorations
Colony	size, shape, attachment and colour; position of branchial, atrial and/or common cloacal apertures; form of the systems and/or arrangement of zooids

Zooid	Size, shape and position of apertures; position of body muscles; position, course and relative size of gut loop and gonads; presence of retractor muscles; lateral organs; vascular appendices
Thorax	shape, width, length and number of rows of stigmata and number/row, presence of papillae on the transverse sinuses, presence of longitudinal vessels or rudiments of them
Gut loop	position, course, subdivisions
Gonads	form, position, size, relationship of ovary to testis, position of embryos
Larvae	size and form of larvae, numbers of adhesive organs and ampullae, blastozooids, sense organs, length of tail, encircling nature of the tail
Spicules	distribution and diameter and form of test spicules by Scanning Electron Microscopy

9.3.3   Invasive Status

In accordance with a terminology adapted from Carlton (1996a, b), identified ascidians were categorized into Cryptogenic, Non-indigenous (NI), Non-indigenous Established (NIE) and Nonindigenous Invasive (NII). The criteria of the WWF (2009) were also used to classify the species.

9.4   Species Composition

The taxonomic coverage of this data set spans the class Ascidiacea. The largest number of records was from the family Didemnidae and Polycitoridae (N = 13), followed by the Polyclinidae (N = 12), Styelidae (N = 6), Pyuridae (N = 5), Ascidiidae (N = 4), Perophoridae (N = 3) and Rhodosomatidae and Molgulidae (N = 1) (Fig. 9.5).

9.5   Taxonomy

Ecteinascidia garstangi Sluiter, 1898 (Plate 9.1a)

Class: Ascidiacea
Suborder: Phlebobranchia
Family: Perophoridae Giard, 1872
Genus: *Ecteinascidia* Herdman, 1880
Species: *Ecteinascidia garstangi* Sluiter, 1898

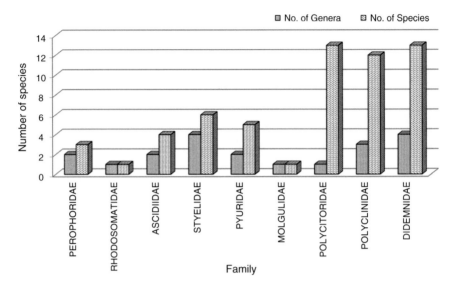

Fig. 9.5 Representation of ascidian fauna belonging to various families

Material Examined Collected from small stones in shallow water at a depth of about half a meter at Chinna Muttom. This species was previously reported from Mandapam by Renganathan (1984).

Distribution Gulf of Mannar, Southeast coast, India.

World Distribution Madagascar, South Pacific Ocean.

Description Colonial ascidian. Zooids are transparent and colourless. Separate small zooids have basal connective and adhered on the substratum. Test is delicate, transparent and free of epibionts. Branchial aperture is terminal. Transverse muscles extend over two-third of the body wall and between atrial and branchial apertures. The gut forms a secondary loop in which gonads are found.

Remarks This species differs from *Ecteinascidia venui, E. thurstoni* by the presence of transverse muscles between branchial and atrial apertures which form the characteristic feature of this species in field and also absence of yellowish orange pigment spots at the base of the siphons. *E. garstangi* contrast with *E. turbinata* by the size of the branchial siphon and also by the absence of orange coloured pigment cells throughout the body. The latter species has longer branchial siphon.

Ecteinascidia venui Meenakshi, 2000 (Plate 9.1b)

Class: Ascidiacea
Suborder: Phlebobranchia
Family: Perophoridae Giard, 1872

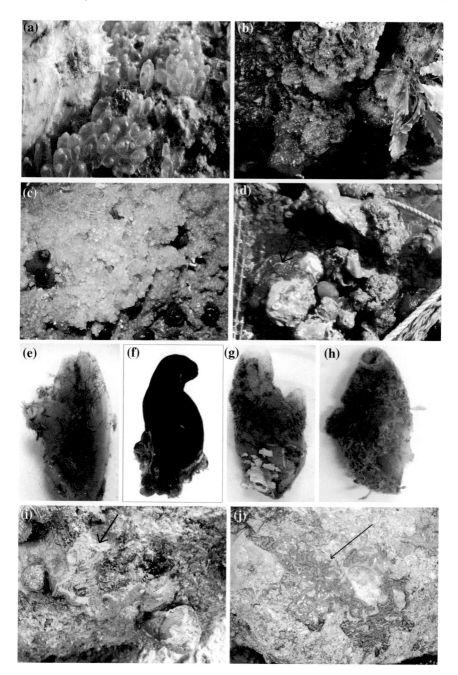

Plate 9.1   a *Ecteinascidia garstangi* (ICM001), **b** *Ecteinascidia venui* (ICM002), **c** *Perophora multiclathrata* (ICM003), **d** *Rhodosoma turcicum* (ICM004), **e** *Phallusia arabica* (ICM005), **f** *Phallusia nigra* (ICM006), **g** *Ascidia gemmata* (ICM007), **h** *Ascidia sydneiensis* (ICM008), **i** *Botryllus purpureus* (ICM009), **j** *Botryllus schlosseri* (ICM010)

Genus: *Ecteinascidia* Herdman, 1880
Species: *Ecteinascidia venui* Meenakshi, 2000

Material Examined Collected from pillars of jetty at a depth of about 1–2 m at Mandapam. This species was previously reported from Thoothukudi coast by Tamilselvi (2008), and Tamilselvi et al. (2011).

Distribution Gulf of Mannar, Southeast coast, India.

Description Zooids are transparent, cylindrical, up to 1.5–2 cm in height with 0.7–0.9 cm wide branchial sac. Large sized zooids are attached to a common branched stolon network with a short stalk at the posterior end of the zooid. Living colonies are light flesh coloured anteriorly with yellowish orange pigment spots on the siphon's rim. The pigment spots cannot be seen while in preservative. The test is thin, transparent and very delicate. Both the branchial and atrial siphons are orange coloured and opens at terminal. Branchial papillae are present. The body wall is thin and transparent with strong circular muscles in the branchial sac and longitudinal muscles in the siphon.

Remarks *E. venui* differs from *E. garstangi* in the absence of transverse muscles between siphons and the presence of yellowish orange band around the rim of apertures but not at the base. This species contrasts with *E. turbinata* in the absence of dark orange coloured pigment cells scattered throughout the body.

Perophora multiclathrata (Sluiter, 1904) (Plate 9.1c)

Class: Ascidiacea
Suborder: Phlebobranchia
Family: Perophoridae Giard, 1872
Genus: *Perophora* Wiegmann, 1835
Species: *Perophora multiclathrata* (Sluiter, 1904)

Material Examined Collected from small stones in shallow water at a depth of about half a meter at Inigo Nagar, Thoothukudi coast and Chinna Muttom, Kanyakumari coast. Previously reported from the coastal waters of Thoothukudi, Bombay, Vizhinjam, Mandapam, Rameshwaram, Valinokkam and Ervadi (Renganathan 1983c; Tamilselvi 2008; Abdul Jaffar Ali et al. 2009; Tamilselvi et al. 2011; Meenakshi 1997).

Distribution Gulf of Mannar, Southeast coast, India.

World Distribution France, North Atlantic Ocean.

Description Zooids are Parrot green in living condition and called "green ascidian". Zooids are found attached to small stones in intertidal flat region. Zooids are 2 mm long and fully matured. There are about 6 transverse muscles in the anterior half of the body along the anterior and posterior sides of the atrial aperture. Five distinct stigmatal rows are observed. The stomach rather elongate (Fig. 9.6).

Fig. 9.6 Whole zooid of
Perophora multiclathrata

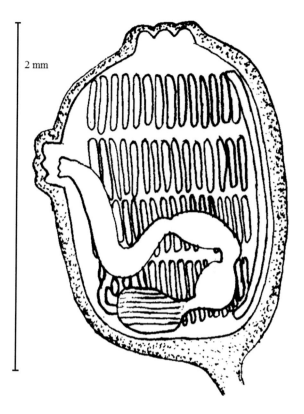

2 mm

Remarks This species is characterized by small zooids with a single horizontal muscle bundle dividing the atrial siphon into two parts. Oka (1931) also observed 6 transverse muscles are identical with present species.

Rhodosoma turcicum (Savigny, 1816) (Plate 9.1d)

   Class: Ascidiacea
   Suborder: Phlebobranchia
   Family: Rhodosomatidae Hartmeyer, 1908
   Genus: *Rhodosoma* Ehrenberg, 1828
   Species: *Rhodosoma turcicum* (Savigny, 1816)

Material Examined This species is encrusting on other ascidians at a depth of about 4–5 m in harbour region at Thoothukudi coast. Previously reported from Thoothukudi coast (Meenakshi 1997).

Distribution Gulf of Mannar, Southeast coast, India.

World Distribution Lebanon, Mediterranean Sea (Eastern Basin), Federal Republic of Somalia, Gulf of Mexico, Mozambique, North Atlantic Ocean.

Fig. 9.7 Whole zooid of
Rhodosoma turcicum

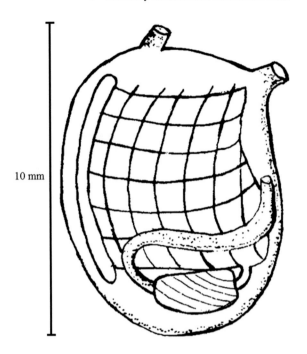

10 mm

Description This simple ascidian grows up to a length from 1 to 5.5 cm. Transparent and yellowish tunic having bright coloured siphons is a notable feature to this species for identification. But the color disappeared in preserved specimens. Each individual is attached to the substrate by the right side of the body. Epibionts adhered on the surface of the transparent test. Numerous conical papillae in the lip of the siphons. Also there are circular and longitudinal fibers in the siphons. A specialized short and thick longitudinal band extends from the siphons to the region of the first intestinal loop. Presence of varied sizes of simple oral tentacles. No papillae in the peripharyngeal groove and the pre-pharyngeal region. The dorsal tubercle is horseshoe shaped and slightly enrolled. The dorsal lamina has long languets. Presence of longitudinal vessels in each side and 4–5 stigmata per mesh. The stomach has longitudinal folds; it is in vertical position. Anus multilobed. Testis is formed by elongated follicles. Parastigmatic vessels and secondary papillae absent (Fig. 9.7).

Remarks Both branchial and atrial siphons are close to each other and specialized muscle fibres present in the siphons. The bright red colour siphon is the characteristic feature of this species for identification.

Phallusia arabica Savigny, 1816 (Plate 9.1e)

   Class: Ascidiacea
   Suborder: Phlebobranchia
   Family: Ascidiidae Herdman, 1882

Genus: *Phallusia* Savigny, 1816
Species: *Phallusia arabica* Savigny, 1816

Material Examined Collected from pearl oyster cages at a depth of about 5–6 m at Thoothukudi Harbour. This species was previously reported from Thoothukudi coast, Vizhinjam Bay (Meenakshi 2003; Tamilselvi 2008; Abdul Jaffar Ali et al. 2009; Tamilselvi et al. 2011).

Distribution Gulf of Mannar, Southeast coast, India.

World Distribution North Atlantic Ocean, Red Sea, South Pacific Ocean.

Description This is also a common large solitary ascidian, length is varied from 3 to 6.8 cm. The test is thick, firmly attached to the substratum and free of epibionts throughout its life. The branchial siphon is long and it opens at terminal whereas the atrial siphon is short and opens at half way at dorsal surface. Well-developed circular and longitudinal muscles are found at both siphons. There are about 9–11 stigmata in a mesh. Primary gut loop is vertical in position and the secondary loop is slightly tilted. It is characterized by the presence of secondary accessory openings in the neural gland duct.

Remarks This species differs from *P. nigra* in its dark grey colour test and more number of stigmata whereas the latter one has dark black colour test.

Phallusia nigra Savigny, 1816 (Plate 9.1f)

Class: Ascidiacea
Suborder: Phlebobranchia
Family: Ascidiidae Herdman, 1882
Genus: *Phallusia* Savigny, 1816
Species: *Phallusia nigra* Savigny, 1816

Material Examined Collected at Thoothukudi Harbour from pearl oyster cages at a depth of about 5–6 m. This species was previously reported from Thoothukudi coast and Vizhinjam Bay, southwest coast of India. (Meenakshi 1997; Tamilselvi 2008; Abdul Jaffar Ali et al. 2009; Tamilselvi et al. 2011).

Distribution Thoothukudi coast, Southeast coast, India.

World Distribution Caribbean Sea, Aegean Sea, Brazil, Greece, Gulf of Mexico, Israel, Mediterranean Sea (Eastern Basin), Panama, Turkey, United Kingdom, United States, Djibouti, North Atlantic Ocean, Red Sea, Venezuela.

Description This is a common large solitary ascidian. It is commonly called as black ascidian in the field. The unique feature is the typical velvety-black or dark-brown colour. Hard Test is free of epibionts throughout its life. Dorsal tubercle is U shaped. Neural duct has accessory openings. Number of stigmata per mesh is about 4–9 and plications in the stomach wall are about 6–7. Posterior intestine is not dilated. Margin of the anus is multilobed. It is characterized by the presence of (16–36) accessory openings in the neural gland duct. Peripharyngeal area with many papillae, longitudinal vessels in pharynx converging to the dorsal lamina in the anterior region. Size of this species can reach up to 10 cm in length.

Remarks Dark black colour test is unique feature of this species commonly found in Port area, whereas the young individual is grey in colour like that of *P. arabica.* Besides fewer numbers of stigmata also differ from the latter species.

Ascidia gemmata Sluiter, 1895 (Plate 9.1g)

    Class: Ascidiacea
    Suborder: Phlebobranchia
    Family: Ascidiidae Herdman, 1882
    Genus: *Ascidia* Linnaeus, 1767
    Species: *Ascidia gemmata* Sluiter, 1895

Material Examined Collected at Thoothukudi Harbour from pearl oyster cages at a depth of about 5–6 m. Previously reported at coastal waters of Madras coast, Thoothukudi coast, Valinokkam, Ervadi, Mandapam, Rameshwaram and Vizhinjam (Krishnan et al. 1989; Meenakshi 1997; Tamilselvi 2008; Abdul Jaffar Ali 2009; Tamilselvi et al. 2011).

Distribution Gulf of Mannar and Vizhinjam bay, Southern India.

World Distribution Central Indo-Pacific.

Description Simple ascidian. Dirty white colour, translucent, slightly wrinkled test with epibionts. No large spherical terminal ampullae on the surface of the test. Well musculature can be seen. Muscles form an irregular mesh on right side. In the longitudinal vessels, no intermediate papillae. Also, lack of tentacular fringe on the lobes of aperture. Posterior position of atrial siphon and visceral mass is the peculiar feature. Number of stigmata is four per mesh. Anus opens to the posterior pole of the gut loop. Intermediate papillae on the longitudinal vessels are absent.

Remarks *A. gemmata* differs from *A. sydneiensis* by its soft test, tentacular fringe on the lobes of aperture and anal border is not lobed. Kott (1997) also observed the irregular mesh on the right side is also observed in our specimen. The species is distinguished by the absence of intermediate papillae in the branchial sac. Absence of intermediate papillae distinguish the present species from *A. thompsoni* (Kott 1952), in which primary branchial papillae are present. Morphologically these species resembles *A. sydneiensis* (Kott 1972) but the test of these species is firmer and thicker.

Ascidia sydneiensis Stimpson, 1855 (Plate 9.1h)

Class: Ascidiacea
Suborder: Phlebobranchia
Family: Ascidiidae Herdman, 1882
Genus: *Ascidia* Linnaeus, 1767
Species: *Ascidia sydneiensis* Stimpson, 1855

Material Examined Collected at Thoothukudi Harbour from pearl oyster cages at a depth of about 5–6 m. Previously reported from Thoothukudi coast (Meenakshi 1998a; Tamilselvi 2008).

Distribution Gulf of Mannar, Southeast coast, India.

World Distribution Brazil, Caribbean Sea, Colombia, Cuba, France, Gulf of California, Israel, Mediterranean Sea—Eastern Basin, Panama, United States, Mozambique, South Africa.

Description Simple ascidian. Translucent, wrinkled and leathery test with epibionts. Prominent short branchial and atrial siphon. Atrial siphon is right angle to branchial siphon. Musculature on the right side of the body. Oral tentacles are simple, and variable in number. There is no papilla in peripharyngeal area. Number of stigmata per mesh is about 6–12. Number of plications in the stomach wall is about 5–9. Dorsal tubercle is U shaped. No secondary accessory openings from the neural duct. Dilated posterior intestine looks sac like pouch. The peculiar feature is the bilobed anus.

Remarks This species differs from *A. gemmata* by its characteristic feature of presence of more number of stigmata, bilobed anus and no irregular mesh on right side of the body. Present species resembles with *A. sydneiensis* (Kott 1972) in the nature of test, branchial and atrial apertures and their positions. In most of the specimens the gut is always filled with mud which appears to be the characteristic of this species.

Botryllus purpureus (Oka, 1932) (Plate 9.1i)

Class: Ascidiacea
Suborder: Stolidobranchia
Family: Styelidae Sluiter, 1895

Subfamily: Botryllinae
Genus: *Botryllus* Gaertner, 1774
Species: *Botryllus purpureus* (Oka, 1932)

Material Examined Reported at Colachel. This species was found in fouling communities on the hull of boat and was collected at a depth of about one meter. Previously reported from Thoothukudi coast, Gulf of Mannar, Southeast coast, India (Meenakshi and Senthamarai 2006).

Distribution Southwest coast, India.

World Distribution North Pacific Ocean.

Description Colonies are large and irregular encrusting sheets fouling on the substratum. Few sand particles distributed on the surface of thin transparent test. Zooids are arranged in a circular system. Colonies are brown, purple or pink in living condition and flesh coloured in preservative. Zooids short, up to 1 mm long with narrow atrial aperture. Terminal branchial aperture has smooth border. Pigment cells scattered throughout the body wall. Branchial sac has 6–7 rows of stigmata with 15–16 stigmata in a row. Atrial aperture opens between the second and fourth row of stigmata. Barrel shaped stomach has 8 obvious folds. Gonads one on each side of the body.

Remarks Presence of sand in the test and unremarkable terminal ampullae and 7 rows of stigmata is distinguished from *B. schlosseri* (Fig. 9.8).

Botryllus schlosseri (Pallas, 1766) (Plate 9.1j)

Class: Ascidiacea
Suborder: Stolidobranchia
Family: Styelidae Sluiter, 1895
Subfamily: Botryllinae
Genus: *Botryllus* Gaertner, 1774
Species: *Botryllus schlosseri* (Pallas, 1766)

Material Examined Reported at Colachel. This species was fouling the hull of boat and the other species of ascidians and was collected at a depth of about one meter. Previously reported from Thoothukudi and Vizhinjam coast, India (Meenakshi 2006a; Abdul Jaffar Ali et al. 2009).

Distribution Southwest coast, India.

Fig. 9.8 Whole zooid of
Botryllus purpureus

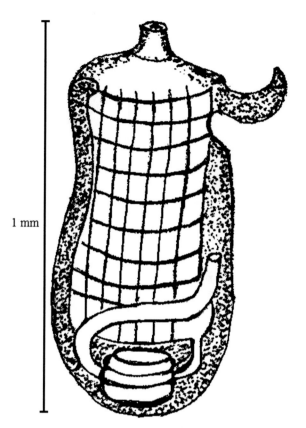

1 mm

World Distribution Adriatic Sea, Arafura Sea, Argentina, Australia, Bay of Fundy, Bismarck Sea, Canada, Chile, Coral Sea, Great Australian Bight, Gulf of Mexico, Japan, Mexico, New Zealand, North Atlantic Ocean, North Pacific Ocean, North Sea, Papua New Guinea, Portugal, Rio de la Plata, Solomon Sea, South Africa, South China Sea, Coastal Waters of Southeast Alaska and British Columbia, United Kingdom, United States, Belgium, France, Mediterranean Sea, Mediterranean Sea (Eastern Basin), Netherlands, Norway.

Description Colonies are encrusting sheets of brightly coloured zooids embedded within a translucent colorless test. Colony up to 5 cm in maximum dimension. 10–15 zooids are arranged in a star shaped or flower shaped system (Star Ascidian). Zooids 1–2 mm long with a smooth border in the branchial aperture. Wide branchial sac with 10–12 rows of stigmata and 20–22 stigmata in a row. Stomach with 10 fine folds. L shaped caecum can be seen in the loop of intestine. Colonies are purple brown in living condition (Fig. 9.9).

Fig. 9.9 Whole zooid of
Botryllus schlosseri

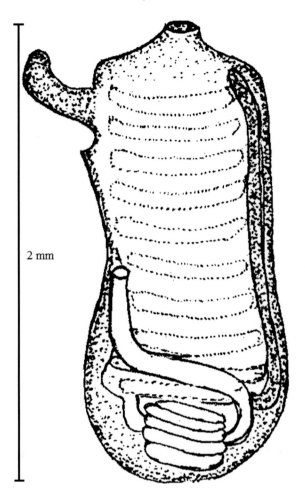

2 mm

Remarks Absence of sand in the test, more number of stigmata per row and rosette shaped male follicles differentiates this species from *B. purpureus*. This species can be distinguished from *B. leachi* by the arrangement of its zooids into star-like systems around common exhalant openings rather than long, meandering systems. The other species which might be confused are the polyclinids *Aplidium nordmani* and *Sidnyum elangans*.

Botrylloides chevalense Herdman, 1906 (Plate 9.2a)

   Class: Ascidiacea
   Suborder: Stolidobranchia
   Family: Styelidae Sluiter, 1895
   Subfamily: Botryllinae

Plate 9.2 a *Botrylloides chevalense* (ICM011), **b** *Botrylloides magnicoecum* (ICM012), **c** *Symplegma oceania* (ICM013), **d** *Styela canopus* (ICM014), **e** *Microcosmus curvus* (ICM015), **f** *Microcosmus exasperatus* (ICM016), **g** *Microcosmus propinquus* (ICM017), **h** *Microcosmum squamiger* (ICM018)

Genus: *Botryllus* Gaertner, 1774

Species: *Botrylloides chevalense* Herdman, 1906

Material Examined Reported at Inigo Nagar. Collected from small stones in shallow water at a depth of about one meter. Previously reported from Thoothukudi coast, Bombay and Goa (Renganathan 1984a; Tamilselvi 2008; Tamilselvi et al. 2011).

Distribution Southern coast, India.

World Distribution Ceylon water, Indian coast.

Description Colonies are encrusting sheets of brightly coloured zooids embedded within a colorless test. Colony up to 10 cm in length. Zooids are arranged in a double rows system. Zooids 1–2 mm long with a smooth border in the branchial aperture. Branchial sac wide with 8–10 rows of stigmata and 20–26 stigmata in a row. Stomach with 10 fine folds. L shaped caecum can be seen in the loop of intestine. Colonies are yellowish brown in living condition.

Botrylloides magnicoecum Hartmeyer, 1912 (Plate 9.2b)

Class: Ascidiacea
Suborder: Stolidobranchia
Family: Styelidae Sluiter, 1895
Subfamily: Botryllinae
Genus: *Botrylloides* Milne-Edwards, 1841
Species: *Botrylloides magnicoecum* Hartmeyer, 1912

Material Examined Reported at Colachel. This species was fouling the hull of boat and encrustiong on other species of ascidians and was collected at a depth of about one meter. Previously reported from Thoothukudi coast, Bombay, Gulf of Mannar, southeast coast, India (Renganathan 1985; Tamilselvi 2008 and Tamilselvi et al. 2011).

Distribution Southeast and southwest coast, India.

World Distribution South Africa, New Zealand.

Description Colonies are thin sheets up to 2–3 cm wide, fouling the substrata. Living colonies are yellow in colour and brownish purple in preservative. Test is thin and transparent. Zooids are arranged in a double row system. Zooids small, up to 1 mm long. Branchial sac is long with 14 rows; each row has 3–4 stigmata. Stomach is short and rounded with 9 folds.

Remarks The double rows of arrangement of zooids are the characteristic feature of this species. The same feature was also observed by Kott (1963). Australian member of this species has described by Kott (1972) are identical with the present colony and zooids especially in record to arrangement of close-set double rows of zooids. *Botrylloides leachi* colonies are similar lobed but the cloacal apertures are present along the side of the head between the double row of zooids.

Symplegma oceania Tokioka, 1961 (Plate 9.2c)

Class: Ascidiacea
Suborder: Stolidobranchia
Family: Styelidae Sluiter, 1895
Subfamily: Polyzoinae
Genus: *Symplegma* Herdman, 1886
Species: *Symplegma oceania* Tokioka, 1961

Material Examined Collected from Thoothukudi harbour fouling the other ascidian species at a depth of about 5–6 m. This species was also collected from the pillars of jetty and hull of boat in Mandapam at a depth of about one meter. Previously reported at Gulf of Mannar by Meenakshi (2003) and Tuticorin by Tamilselvi (2008) and Tamilselvi et al. (2011).

Distribution Gulf of Mannar, Southeast coast, India.

World Distribution Indonesia, Hong Kong, China (People's Republic), Palau, New Caledonia, Fiji, Sri Lanka, Northern Territory (N coast), Queensland (Central East coast, Great Barrier Reef, Northeast coast), South Australia (Great Australian Bight, South Gulfs coast), Victoria (Bass Strait), Western Australia (Central West coast, Lower West coast, Northwest coast); Noumea, also West Indian Ocean, Gulf of Mexico, Caribbean.

Description The species forms flattened colonies of dome shaped zooids with three colour morphs; grey, pink and yellow. Test is somewhat rough in which zooids are arranged loosely. Seldom 4 branchial folds. 8 to 14 rows of stigmata. No cloacal system. Pair of gonads is present on either side of the gut. Hook like structure is present at the base of the stomach. There are about 10–16 folds in the stomach. The arrangement of the gastro-intestinal duct and vessels is extremely variable and includes the arrangements described by Tokioka for *S. oceania* and for *S. viride*. The single or branched duct extends from a variable level between the middle and base of the outer convex side of the gastric caecum (which may be curved or almost straight) to the descending limb of the primary gut loop. Single or branched vessels also extend from the tip of the caecum to the ascending limb of the primary gut loop (distal to the stomach) where they ramify over the intestinal wall. These vessels and

Fig. 9.10 Whole zooid of
Symplegma oceania

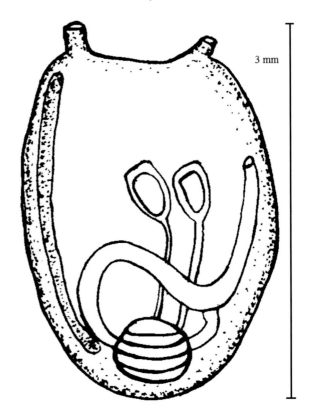

3 mm

ducts associated with the caecum are very delicate, embedded in the membranes of
the body wall that cover the gut loop. Gonads are large and few (Fig. 9.10).

Remarks Species forms flattened colonies of dome shaped zooids with two colour
morphs, pink and yellow, encrusting on the available substrata. Larva is black in
colour, dome shaped with short straight but slightly curved at the posterior end is
the characteristic feature. Three intestinal connectives from pyloric caecum
resemble the character of this species. This species resembles *Symplegma reptans*
and *S. viride* (Kott 1964) in number of stigmata but differs in pyloric caecum
connectives. *S. brakenhielmi* also has 10 rows of stigmata but two intestinal connectives from pyloric caecum.

Styela canopus (Savigny, 1816) (Plate 9.2d)

    Class: Ascidiacea
    Suborder: Stolidobranchia
    Family: Styelidae Sluiter, 1895
    Subfamily: Styelinae

Genus: *Styela* Fleming, 1822
Species: *Styela canopus* (Savigny, 1816)

Material Examined Reported at Thoothukudi harbour, Inigo Nagar, Muthunagar beach, Tiruchendur, Chinna Muttom, Colachel and Mandapam. This species was found attached with molluscan shells and were collected at a depth of 5–6 m in harbour region, in the pillars of jetty at one meter depth and also in the shallow water regions. Previously reported from Thoothukudi coast, Bombay and Vizhinjam (Renganathan 1986b; Meenakshi 1997; Tamilselvi 2008; Abdul Jaffar Ali et al. 2009 and Tamilselvi et al. 2011).

Distribution Gulf of Mannar, Southeast coast, India.

World Distribution Japan, France, Indonesia, Hong Kong, New South Wales (Central East coast), Queensland (Central East coast, Northeast coast), Western Australia (Lower West coast, Northwest coast); Torres Strait, Coral Sea, west Pacific Ocean, Korea, tropical and temperate Atlantic Ocean, Persian Gulf, Adriatic, Mediterranean, Ascension Is., Channel Is., West coast of France.

Description This simple ascidian is readily identified by its deeply curved gut loop, long stomach with fine longitudinal pleats, long anal lobes, and leaf-like endocarps on and in the gut loop. White leathery test. The test is firmly attached with the mantle body.

Remarks The present specimen resembles the previously recorded species described by Savigny (1816) and Kott (1985). The test is tough, leathery fleshy white coloured in living condition. Presence of brown coloured wrinkles around the lobed branchial siphon is a peculiar feature in field identification.

Microcosmus curvus Tokioka, 1954 (Plate 9.2e)

Class: Ascidiacea
Suborder: Stolidobranchia
Family: Pyuridae Hartmeyer, 1908
Genus: *Microcosmus* Heller, 1877
Species: *Microcosmus curvus* Tokioka, 1954

Material Examined Collected at Muthunagar Beach and Chinna Muttom from small stones in shallow water at a depth of one meter. Previously reported from Thoothukudi coast (Renganathan 1983a; Meenakshi 1997; Tamilselvi 2008; Tamilselvi et al. 2011).

Distribution Gulf of Mannar, Southeast coast, India.

World Distribution Wake Island, Palau, Queensland (Great Barrier Reef); Marianas Is., Tokara Is., Tahiti, Indian Ocean.

Description The smallest simple ascidian in *Microcosmus* genus. The branchial aperture is terminal, directed downwards, and the atrial aperture from the antero—dorsal surface, is directed obliquely upwards. Irregular projections of the surface are present on the anterior part of the body. The siphonal lining has minute, overlapping, flattened, pointed spines. The body wall is strong and muscular. The opening of the neural gland on the dorsal tubercle forms a single coil about one and a half times in a clockwise direction from the center. Its distinctive characters are smaller siphonal spines, rudimentary sevenths fold usually present only on the left side, undulating and curving gonads with gonoducts directed anteriorly. The species are light yellow, light blue or light red in living condition.

Remarks *M. curvus* (Tokioka 1954) described by Tokioka 1954 in Japan and in Australia by Kott (1969). Renganathan (1983a) first reported in the Gulf of Mannar, Southeast coast of India, and recorded largest specimen measures 15 mm in length. They are either violet or red in colour. The species characteristics are its six branchial folds on each side, absence of siphonal armature, four tough cartilaginous spoon-shaped valves at the base of the branchial siphon and a tough, fibrous atrial velum divided into two and sometimes forming pocket valves. The body wall is strong and muscular.

Microcosmus exasperatus Heller, 1878 (Plate 9.2f)

   Class: Ascidiacea
   Suborder: Stolidobranchia
   Family: Pyuridae Hartmeyer, 1908
   Genus: *Microcosmus* Heller, 1877
   Species: *Microcosmus exasperatus* Heller, 1878

Material Examined This species was collected at Thoothukudi harbour from oyster cages at a depth of about 5–6 m also found attached to the pillars of jetty at one meter depth in Mandapam. Previously reported from Thoothukudi coast (Krishnan 1989; Meenakshi 1997; Tamilselvi 2008; Abdul Jaffar Ali 2009; Tamilselvi et al. 2012).

Distribution Gulf of Mannar, Southeast coast, India.

World Distribution Fiji, Philippines, Hawaii, New Caledonia, Bermuda, Brazil, Indonesia, Florida, New South Wales (Central East coast, Lower East coast), Northern Territory (North coast), Queensland (Central East coast, Northeast coast), Western Australia (Central West coast, Lower West coast, North coast, Northwest coast); West Indies, East Africa, Red Sea.

Description The globular body is enclosed within a leathery and wrinkled tunic. The tunic is orange or purple, maintaining colour in formalin, and contains some sand and encrusting organisms on the surface. Both siphons are lobed with four triangular lobes. There are 12 large and 18 smaller branched oral tentacles arranged on a muscular ring. The prepharyngeal groove is double, the neural gland opening U-shaped with much in rolled horns. The pharynx has 8 folds on each side, but the ventral fold on the right side is incomplete. One gonad on each side of the body is formed by 3–5 masses of testis follicles surrounding a tubular ovary. Branchial siphon has numerous siphonal spines, characteristic feature of this species. The species are light brown, light yellowish brown and light radish brown in living condition.

Remarks *Microcosmus helleri* differs from *M. exasperatus* by the smaller number of branchial folds, the gut loop is less bent upwards at the anterior end and there is a more compacted gonad, not divided into segments (Van Name 1945). *Microcosmus squamiger* has the gonads also divided into three lobes and has been confused with *M. exasperatus*, but the siphonal spines are very different, with roof-tile shape and spiny rims (Monniot et al. 2001). *Microcosmus exasperatus* is widely distributed in all oceans and, in the Palk Strait and Gulf of Mannar regions, its Southeast Coast of India distributional limits (Meenakshi 1997, 2005, 2008). But in the present report *M. exasperatus,* a new report in Gopalapattinam Coast, Palk Bay region. Nobody surveyed this area.

Microcosmus propinquus Herdman, 1882 (Plate 9.2g)

> Class: Ascidiacea
> Suborder: Stolidobranchia
> Family: Pyuridae Hartmeyer, 1908
> Genus: *Microcosmus* Heller, 1877
> Species: *Microcosmus propinquus* Herdman, 1882

Material Examined This species was collected at Thoothukudi harbour from oyster cages at a depth of about 5–6 m also found attached to the pillars of jetty at one meter depth in Mandapam. Previously reported from Thoothukudi coast (Meenakshi 2003).

Distribution Gulf of Mannar, Southeast coast, India.

World Distribution Australia.

Description The animal is roughly round, attached with posterio-ventral part of the body to the available substratum. Branchial siphon at the anterior end and the atrial at middle with four lobed. Test is leathery and smooth on surface with orange coloured siphons. Inner surface of the siphons armed with both spines and scales. The mantle body is very soft with thin mantle, nearly translucent and yellowish in colour. The siphons are thick and orange in colour. The musculature is regular consisting of longitudinal muscles covering branchial siphons and transverse ones diverging from the atrial siphon. Branchial folds 7 on each side. Parastigmatic vessel present. And about 14 stigmata in a mesh. Peribranchial area is papillated.

Remarks The present specimen resembles the described specimens of *M. propinquus* described by Meenakshi (1997). The number of stigmata in a mesh and the presence of both spines and scales in the inner lining of branchial siphon seems to be closely related to the specimen described by Nishikawa and Tokioka (1976).

Microcosmus squamiger Michaelsen, 1927 (Plate 9.2h)

> Class: Ascidiacea
> Suborder: Stolidobranchia
> Family: Pyuridae Hartmeyer, 1908
> Genus: *Microcosmus* Heller, 1877
> Species: *Microcosmus squamiger* Michaelsen, 1927

Material Examined This species was collected from oyster cages in Thoothukudi harbour at a depth of about 5–6 m. Previously reported from Thoothukudi coast (Meenakshi and Senthamarai 2007; Tamilselvi 2008; Tamilselvi et al. 2012) and in Royapuram coast of Chennai (Krishnan 1952).

Distribution Gulf of Mannar, Southeast coast, India.

World Distribution New South Wales (Central East coast, Lower East coast), Queensland (Central East coast, Northeast coast), South Australia (S Gulfs coast), Tasmania (Bass Strait, Tas. coast), Victoria (Bass Strait), Western Australia (Central West coast, Lower West coast, Southwest coast), Tyrrhenian Sea, Italy, Western Indian Ocean, Mediterranean Sea,. Red sea, India.

Description *M. squamiger* is a solitary ascidian attached on the hulls of ship and oyster bed. The species are light brown or light yellowish brown in living condition. In its habitat, the 5–6 individuals of different sizes can be seen. The tough test provides the space for attachment for many epibionts. The gonads are divided into three lobes and has been confused with *M. exasperatus*, but the siphonal spines are very different, with roof-tile shape and spiny rims (Monniot and Monniot 2001).

Remarks The specimen studied resembles *M. squamiger* (Michaelsen 1927) and Kott (1985) in many features. This species is very often confused with the similar species of *M. exasperatus;* except the shape of the spines in branchial siphon. In the former species, scales are found whereas in the latter, pointed spines in the siphon.

Herdmania momus (Savigny, 1816) (Plate 9.3a)

    Class: Ascidiacea
    Suborder: Stolidobranchia
    Family: Pyuridae Hartmeyer, 1908
    Genus: *Herdmania* Lahille, 1888
    Species: *Herdmania momus* (Savigny, 1816)

Material Examined This species was collected from oyster cages in Thoothukudi harbour at a depth of about 5–6 m and also found in large numbers in nearby stations such as Kayalpattinam and Veerapandianpattinam at a depth of 25 m for the first time. Previously reported from Thoothukudi coast (Das 1936; Sebastian 1952; Renganathan 1983; Tamilselvi 2008; Tamilselvi et al. 2011, 2014, 2015), Valinokkam and Krusadai Island.

Distribution : Gulf of Mannar, Southeast coast, India.

World Distribution Japan, Taiwan, Philippines, Indonesia, Palau, Hawaii, Fiji, South Africa, Queensland (Northeast coast), Western Australia (Central West coast, Lower West coast, Northwest coast); also west Indian Ocean, south China Sea, Tahiti, Arafura Sea.

Description *Herdmania momus* is a solitary ascidian. The test is tough and leathery in nature, provides a habitat for a number of epibionts. The acicular spicules of the body wall concentrate mainly between the siphons and in the ventral region. There are up to 15 wide and branched oral tentacles. The prepharyngeal groove has a double margin and forms a deep V around the dorsal tubericle, whose aperture is horseshoe-shaped with spiraled ends. The dorsal lamina is subdivided into several thin languets. The pharynx has eight folds on each side of the body, with longitudinal vessels that vary from 8–15 on each folds. Parastigmatic vessels are present. The esophagus is short and the digestive gland of the stomach has a small portion on the right side and a larger portion on the left side of the stomach. The intestine is attached

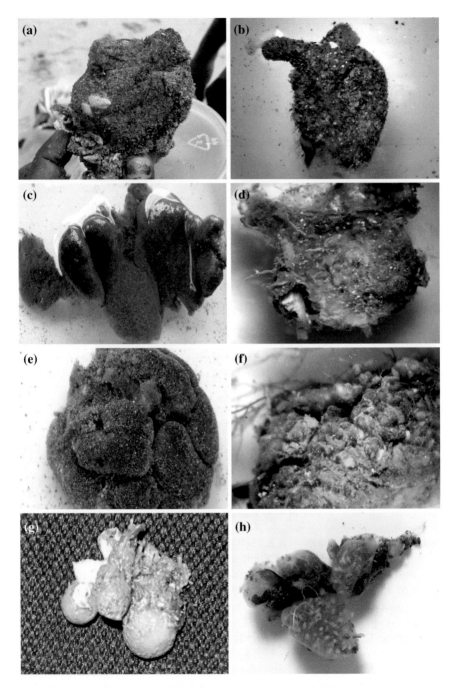

Plate 9.3 a *Herdmania momus* (ICM019), **b** *Molgula ficus* (ICM020), **c** *Eudistoma amplum* (ICM021), **d** *Eudistoma carnosum* (ICM022), **e** *Eudistoma constrictum* (ICM023), **f** *Eudistoma gilboviride* (ICM024), **g** *Eudistoma lakshmiani* (ICM025), **h** *Eudistoma laysani* (ICM026)

to the body wall with a closed primary loop and an open and small secondary loop. The anus is slightly lobed. The gonad on the right side is ventral. Gut loop alongside pharynx; body not divided. Gonads usually on both sides of the body; brachial sac always folded. Tentacles branched. Stigmata straight. Dorsal lamina a series of languet. Barbed spines are numerous in both test and body, characteristic feature of this species. The species are pink or light red in living condition.

Remarks The species shows great range of colour variations. Underwater photos of the present specimens show red colour. This Genus was first described in India by Das (1936, 1945) and in Madras coast by Sebastian and Kurian (1981). Recently Kartikeyan et al. (2009) described in Palk Bay, Southeast coast of India. Presence of barbed spines in the test and mantle bodies are the characteristic feature.

Molgula ficus (Macdonald, 1859) (Plate 9.3b)

Class: Ascidiacea
Suborder: Stolidobranchia
Family: Molgulidae Lacaze-Duthiers, 1877
Genus: *Molgula* Forbes, 1848
Species: *Molgula ficus* (Macdonald, 1859)

Material Examined This species was collected at Tiruchendur from a large embedded rock at one meter depth. Previously reported by Meenakshi (2003).

Distribution Gulf of Mannar, Southeast coast, India.

World Distribution Chile, United States, Central Indo-Pacific.

Description It is a globular, vase-like simple ascidian with both siphons on the upper surface. The tunic is fleshy white in colour, leathery texture, closely attached to the mantle body. Usually mud adhering on the test. The branchial siphon has six lobes, whereas the atrial siphon has four. Two rows of small pointed papillae, in the fringed siphons.

Remarks *M. ficus* differs from *M. clavata* in apertures. In the latter species, siphons are sessile externally but are present internally. The presence of 6 and 4 pointed lobes in the branchial and atrial apertures respectively, resembles *Molgula sabulosa* (Kott 1975) and differs from the absence of hard and brittle test with hairs.

Eudistoma amplum (Sluiter, 1909) (Plate 9.3c)

Class: Ascidiacea
Suborder: Aplousobranchia
Family: Polycitoridae Michaelsen, 1904

Genus: *Eudistoma* Caullery, 1909
Species: *Eudistoma amplum* (Sluiter, 1909)

Material Examined This species was collected at Tiruchendur from a large embedded rock at one meter depth. Previously reported in Gulf of Mannar by Meenakshi (2003).

Distribution Gulf of Mannar, Southeast coast, India.

World Distribution Tanzania, Central Indo-Pacific.

Description Colonies are robust, irregular flat topped cushions up to 10 cm or more in maximum dimension and about 1.5 cm thick with curved borders. The surface test is smooth and free from sand grains, but very small sand particles can be seen around the zooids. Zooids are arranged in a circular system. The test is soft and transparent. Zooids are short, 5 mm long and white in preservative. The atrial siphon is longer than that of the branchial siphon. The stomach is at the posterior end of the abdomen and is small and smooth (Fig. 9.11).

Remarks The distinctive feature of this species is extensive sheet like colonies, embedded sand and plant cells in the colourless test. Also a large larva is incubated. The species is distinguished by the presence of number of circular green cells and sand embedded in the colourless test. *Eudistoma vitatum* contains the same test inclusions as the present species but it is distinguished by its relatively small upright lobes with common basal membrane. Zooids of *Polycitor discolor* also identical with those of the present species.

Eudistoma carnosum Kott, 1990 (Plate 9.3d)

Class: Ascidiacea
Suborder: Aplousobranchia
Family: Polycitoridae Michaelsen, 1904
Genus: *Eudistoma* Caullery 1909
Species: *Eudistoma carnosum* Kott, 1990

Material Examined This species was collected at Mandapam from the hull of boat at one meter depth.

Distribution New Record. Gulf of Mannar, Southeast coast, India.

World Distribution Australia.

Fig. 9.11 Whole zooid of
Eudistoma amplum

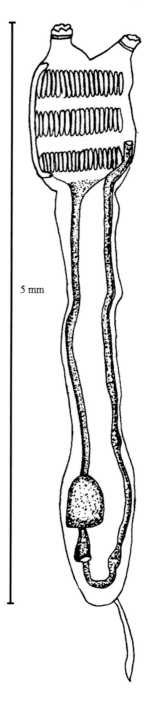

5 mm

Description Colony flat cushion shaped with cloacal apertures on conical promi-
nence. Light greenish in colour with sand sparsely distributed on surface. Zooid is
short up to 6 mm long. Number of stigmata is 16 per row. Stolonic vessels present
and are branched. Longitudinal muscles of up to 27 bands extended up to stomach.
Stomach is posterior. Male follicles are numerous up to 20 with one matured ova
(Fig. 9.12).

Fig. 9.12 Whole zooid of
Eudistoma carnosum

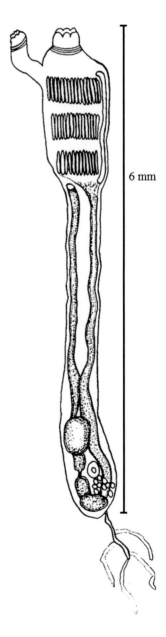

6 mm

Remarks The present species is also similar to *E. ovatum* in the presence of sharp ectodermal processes around the apertures. Large cloacal cavities in their obvious sand-free area of test are characteristic features of this species.

Eudistoma constrictum Kott, 1990 (Plate 9.3e)

Class: Ascidiacea
Suborder: Aplousobranchia
Family: Polycitoridae Michaelsen, 1904
Genus: *Eudistoma* Caullery, 1909
Species: *Eudistoma constrictum* Kott, 1990

Material Examined This species was collected at Tiruchendur (8°29′49″N·78°7′ 32″E) from a large embedded rock at one meter depth. Previously reported from Mandapam (Meenakshi 2002).

Distribution Gulf of Mannar, Southeast coast, India.

World Distribution North Indian Ocean.

Description Colonies are irregular cushions with 3–4 cm in maximum extent and up to 1 cm high. Sand as well as faecal pellets embedded throughout the upper surface of the test and zooids open separately in the sand free area. Sand particles are relatively less at the base. Zooids are arranged in a compartment in the test. A narrow constriction between the thorax and abdomen can be seen. Zooids 4–5 mm in size and are light yellow in preservative. Longitudinal muscles extend into the abdomen as bands. Both branchial and atrial siphons are very short. Zooids have 3 rows of about 12–14 stigmata. A long stolonic vessel extends from the abdomen (Fig. 9.13).

Remarks The peculiar features of this species are the presence of sand and faecal pellets in the test. The zooid is found in a compartment in the test. Presence of narrow constriction between the thorax and abdomen. Absence of circular systems and colony is not lobed.

Eudistoma gilboviride (Sluiter, 1909) (Plate 9.3f)

Class: Ascidiacea
Suborder: Aplousobranchia
Family: Polycitoridae Michaelsen, 1904
Genus: *Eudistoma* Caullery, 1909
Species: *Eudistoma gilboviride* (Sluiter, 1909)

Fig. 9.13 Whole zooid of
Eudistoma constrictum

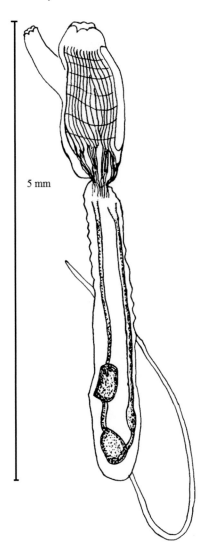

5 mm

Material Examined Reported at Mandapam, Thoothukudi Harbour, Hare Island, North Break Water, Inigo Nagar, Muthunagar beach and Tiruchendur. This species was collected from submerged rocks at a depth of 1–2 m and also from shallow water regions. Previously reported in Gulf of Mannar by Meenakshi (2003).

Distribution Gulf of Mannar, Southeast coast, India.

World Distribution Central Indo-Pacific.

Description The colony is lobed, free of sand on the upper surface of the lobes. The size of the colony is 5–6 cm long and 2–2.5 cm in width. Colony is wide, translucent with sparsely distributed sand particles. Entire base of the colony is attached to the substrates such as sponges, rocks and pillars of the jetty. Zooid is 3–4 mm in size. Branchial siphon upright in position. Atrial siphon is longer than the branchial siphon. Thorax is narrow with 3 rows of stigmata and 16 stigmata per row. Long longitudinal meshes like muscles present till the end of the abdomen. Long stolonic muscles are present. Zooids have long esophageal neck with small stomach. 10–12 male follicles and 1 ovum are present. 4 embryos are present in atrial cavity (Fig. 9.14).

Remarks The wedge shaped, flat topped colonies and dark pigment cells on the surface of the test are the distinctive features of this species. Kott (1990) observed the presence of green cells in *Eudistoma gilbovirde* but in the present specimens the cells are not observed. In *E. glaucum* numerous lobes arise from common basal test but they are different colour and are larger and more rounded.

Eudistoma lakshmiani Renganathan, 1986 (Plate 9.3g)

Class: Ascidiacea
Suborder: Aplousobranchia
Family: Polycitoridae Michaelsen, 1904
Genus: *Eudistoma* Caullery, 1909
Species: *Eudistoma lakshmiani* Renganathan, 1986

Material Examined This species was collected at Inigo Nagar and Muthunagar Beach from small stones at one meter depth. Previously reported from Thoothukudi coast (Renganathan 1986; Tamilselvi 2008).

Distribution Gulf of Mannar, Southeast coast, India.

Description Colonial ascidian. Colonies are lobed. The lobes are smaller having rounded head, covered by delicate transparent test. Siphons are slightly extending; no pigment spot in the siphon. The branchial sac has three rows of rectangle stigmata. The abdomen is longer than the thorax. Gonads are in the gut loop. The unique features of this species are the existence of lobed colonies with a common test devoid of sand.

Remarks Stalked colonies with sand embedded in stalk. Presence of four pairs of ampullae and subdivided into two at the distal end of the larvae are the unique features.

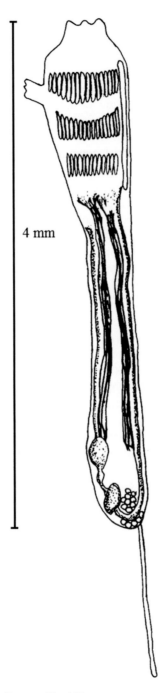

Fig. 9.14 Whole zooid of *Eudistoma gilboviride*

Eudistoma laysani (Sluiter, 1900) (Plate 9.3h)

    Class: Ascidiacea
    Suborder: Aplousobranchia
    Family: Polycitoridae Michaelsen, 1904
    Genus: *Eudistoma* Caullery, 1909
    Species: *Eudistoma laysani* (Sluiter, 1900)

Material Examined Colonies of this species were collected at Inigo Nagar from small stones at a depth of one meter. Previously reported from Mandapam (Meenakshi 2002).

Distribution Gulf of Mannar, Southeast coast, India.

World Distribution Pacific Ocean and Indian Ocean.

Description Colonial ascidian. Colonies are lobed. The lobes are smaller having rounded head, covered by delicate transparent test. The crowded zooids made the colonies rigid and have not in systems. Siphons are slightly extending; no pigment spot in the siphon. The branchial sac has three rows of rectangle stigmata. The atrial siphon originates opposite to the first row of stigmata. The abdomen is longer than the thorax. Gonads are in the gut loop. The unique features of this species are the existence of lobed colonies with a common test devoid of sand and the terminal part of the lobe with bluish white iridescence.

Remarks The present species is similar to that of *Eudistoma tumidum* in majority of features except the lobed colonies with the zooids in circles and differs from *E. lakshmiani* in small zooids, stalked colonies without embedded sand in the stalk.

Eudistoma microlarvum Kott, 1990 (Plate 9.4a)

    Class: Ascidiacea
    Suborder: Aplousobranchia
    Family: Polycitoridae Michaelsen, 1904
    Genus: *Eudistoma* Caullery, 1909
    Species: *Eudistoma microlarvum* Kott, 1990

Material Examined This species was reported at Tiruchendur. Several colonies of this species were found attached to large embedded rock and were collected at a depth of one meter. Previously reported from Mandapam (Meenakshi 2003).

Distribution Gulf of Mannar, Southeast coast, India.

World Distribution South Pacific Ocean.

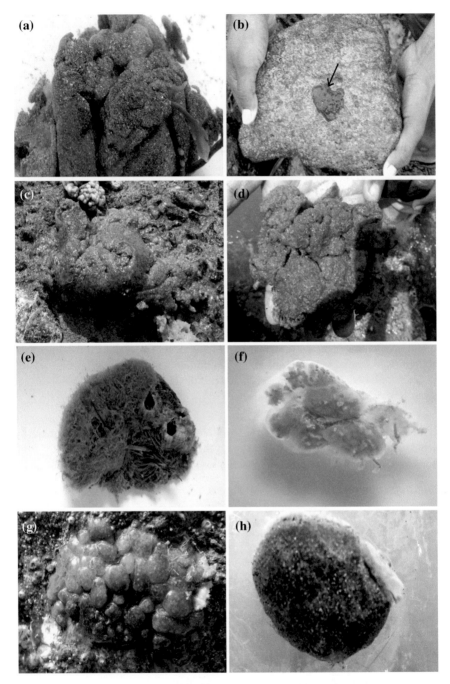

Plate 9.4 **a** *Eudistoma microlarvum* (ICM027), **b** *Eudistoma muscosum* (ICM028), **c** *Eudistoma ovatum* (ICM029), **d** *Eudistoma pyriforme* (ICM030), **e** *Eudistoma reginum* (ICM031), **f** *Eudistoma tumidum* (ICM032), **g** *Eudistoma viride* (ICM033), **h** *Synoicum citrum* (ICM034)

Description The colony is divided into irregular lobes up to 2.0 cm high and 1.5 thick. Sand embedded throughout the surface test, sparsely distributed at lower half of the colony. Test is transparent. Zooids are very small up to 4 mm long when extended and are white and thread-like in preservative. They do not form systems. Both branchial and atrial siphons are relatively short. The branchial sac has 3 rows of up to 8 stigmata (Fig. 9.15).

Remarks The present species resembles the colony of *E. constrictum* for lacks systems but differs in thread like zooids for the former one. *E. globosum, E. elongatum* and *E. laysani* has also sand and lacks system.

Eudistoma muscosum Kott, 1990 (Plate 9.4b)
Class: Ascidiacea
Suborder: Aplousobranchia
Family: Polycitoridae Michaelsen, 1904
Genus: *Eudistoma* Caullery, 1909
Species: *Eudistoma muscosum* Kott, 1990

Material Examined This species was reported at Hare Island and North Break Water. Colonies of this species were collected from small stones at a depth of one meter. Previously reported from Mandapam (Meenakshi 2003).

Distribution Gulf of Mannar, Southeast coast, India.

World Distribution Australia and India.

Description Colonies are smooth and rounded cushions about 1.0 to 2.0 cm in maximum extent. Brown colour in preservative. Zooids are arranged in a circular system. Sand and faecal pellets are embedded in test at the base of the colony. Zooids are large up to 5 mm long and robust. In preserved zooids the endostyle, gut and gonads become dark brown in colour. Dark capsular pigment present in between the branchial and atrial siphons. The oesophageal neck is relatively thick than other *Eudistoma* species. The atrial siphon is three folds longer than branchial siphon. Sphincter muscles surround the siphons. There are numerous thoracic longitudinal and circular muscles in thorax. About 20 stigmata in each row. Longitudinal bands extend up to the abdomen. One or two embryos are observed in the atrial cavity (Fig. 9.16).

Remarks Living specimens are distinguished by their dark green colour. Presence of single capsular pigment is the characteristic feature of this species when compared to *Eudistoma viride* where two endostylar pigments present. Likewise, lack of cloacal cavities in this species differ from *E. reginum* where a lobed cloacal aperture over a rudimentary cloacal cavity.

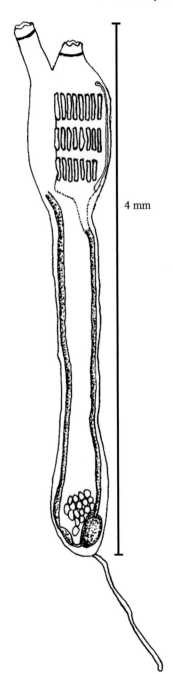

4 mm

Fig. 9.15 Whole zooid of *Eudistoma microlarvum*

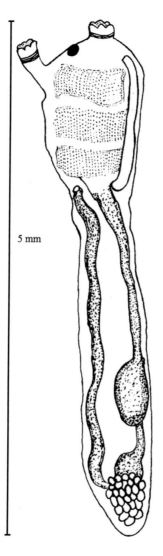

Fig. 9.16 Whole zooid of *Eudistoma muscosum*

Eudistoma ovatum (Herdman, 1886) (Plate 9.4c)

    Class: Ascidiacea
    Suborder: Aplousobranchia
    Family: Polycitoridae Michaelsen, 1904
    Genus: *Eudistoma* Caullery, 1909
    Species: *Eudistoma ovatum* (Herdman, 1886)

Fig. 9.17 Whole zooid of
Eudistoma ovatum

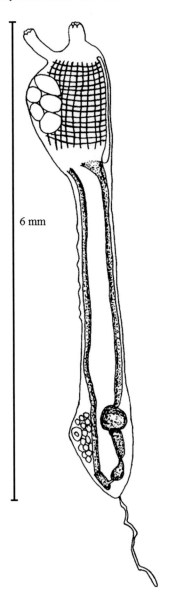

6 mm

Material Examined A single colony of this species was collected at Mandapam from hull of boat at a depth of two meters. Previously reported from Mandapam (Meenakshi 2002).

Distribution Gulf of Mannar, Southeast coast, India.

World Distribution North Indian Ocean.

Description The colony is flat, thick encrusting sheet with 3–4 cm width and 1 cm thickness. Colonies without lobes and sand embedded throughout the colony. The test is transparent and the zooids are not crowded. Zooids pink with a brownish stomach, distributed evenly in the test. Faecal pellets are also distributed in the test. Colony is light brown in preservative. Zooids are short linear and 6 mm long. The branchial and atrial siphons are conspicuous and muscular. Longitudinal and transverse vessels are present throughout the zooid. 3 rows of about 14–15 stigmata. Gonads and stomach are in the posterior end of the abdomen (Fig. 9.17).

Remarks Thin coat of sand with interruption at rounded lobes of apertures. Even though, some features are similar to *E. constrictum* such as the presence of sand and faecal pellets in the test but zooids not found in compartment. The spiral in the gut loop and the inclusions of various particles in the test are characteristic of this specimen. Kott (2008) also described the same characters. *Polycitor multiperforatus* Sluiter (1909) also a similar species although it appears to have more stigmata per row (30) and the circular arrangement of zooids was not detected in the syntype (Kott 1990a). *E. volgare* has similar colony, zooids and larvae to those of the present species appears a synonym (Monniot 1998).

Eudistoma pyriforme (Herdman, 1886) (Plate 9.4d)

> Class: Ascidiacea
> Suborder: Aplousobranchia
> Family: Polycitoridae Michaelsen, 1904
> Genus: *Eudistoma* Caullery, 1909
> Species: *Eudistoma pyriforme* (Herdman, 1886)

Material Examined Colonies of this species were collected at Mandapam from the pillars of jetty at a depth of two meters. Previously reported from Thoothukudi coast (Meenakshi 2003).

Distribution Gulf of Mannar, Southeast coast, India.

World Distribution Madagascar, Central Indo-Pacific.

Description Colonies are lobed and robust with sand throughout the surface test. Common cloacal openings are absent. Colonies are 10 cm long and 1.5–2 cm thick. Zooids are linear and 2–3 mm in length. Short thorax with 3 rows of stigmata. Pigment cells distributed throughout the thorax. A long stolonic vessel at the end of the abdomen (Fig. 9.18).

Fig. 9.18 Whole zooid of
Eudistoma pyriforme

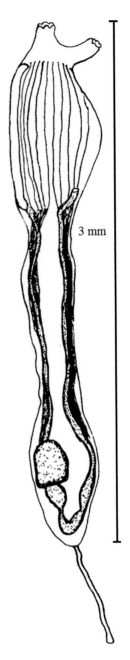

Remarks Separate lobes in the present species deviate from other sandy species of
Eudistoma. Though it resembles *E. globosam* at some extent, its rounded lobes and
crowded zooids differs from the present species.

Eudistoma reginum Kott, 1990 (Plate 9.4e)

Class: Ascidiacea
Suborder: Aplousobranchia
Family: Polycitoridae Michaelsen, 1904
Genus: *Eudistoma* Caullery, 1909
Species: *Eudistoma reginum* Kott, 1990

Material Examined A single colony of this species was collected at Mandapam from hull of boat at a depth of two meters. Previously reported from Mandapam (Meenakshi 2003).

Distribution Gulf of Mannar, Southeast coast, India.

World Distribution West Central Pacific.

Description Colony is flat topped cushion up to 5 cm in maximum dimension, 3 cm wide and about 1.5 cm thick. Whole bottom of the colony is attached to the substrate forming a thick sheet like structure. Colony with several lobes with 7–8 zooids arranged in a circular system. The colony is free of sand grains but thin sandy layer present on gelatinous mass over the entire colony. Test is gelatinous and transparent. Zooids are large and ranging from 0.5 to 1.0 cm long and are white in preservative. Pigment cells present in the test surrounding the zooids. Large vesicles, up to 0.3 mm, are present at all levels in the test, particularly conspicuous (but never crowded together) at the upper surface between zooid openings. Atrial siphon short. Several small male follicles on the left side of the gut loop and a large ovum at the right.

Remarks The present species differs from *E. muscosum* in its smaller cloacal system, rudimentary cloacal cavity, short siphons, and large vesicles in the test. It resembles the latter one by the presence of dark pigment in the posterior horns of the haemocoel.

Eudistoma tumidum Kott, 1990 (Plate 9.4f)
Class: Ascidiacea
Suborder: Aplousobranchia
Family: Polycitoridae Michaelsen, 1904
Genus: *Eudistoma* Caullery, 1909
Species: *Eudistoma tumidum* Kott, 1990

Material Examined A single colony of this species was collected at Mandapam from hull of boat at a depth of two meters. Previously reported from Mandapam (Meenakshi 2003).

Distribution Gulf of Mannar, Southeast coast, India.

World Distribution Central Indo-Pacific.

Description Colony is lobed, translucent and light green in colour attached from single common base. Upper surface of the each lobe is wide. Bottom of the lobe is sparsely sandy. Zooid is very short up to 5 mm long. Longitudinal muscles are mesh like and extend up to stomach. 16–18 male follicles and 4 ova are present (Fig. 9.19).

Remarks The present species is characterized by its naked, lobed colony with the zooids in circles. *E. laysani* also has the same characteristic features but the smaller lobes, zooids and larvae differentiate from the former one. Besides, zooids are not in systems.

Eudistoma viride Tokioka, 1955 (Plate 9.4g)

    Class: Ascidiacea
    Suborder: Aplousobranchia
    Family: Polycitoridae Michaelsen, 1904
    Genus: *Eudistoma* Caullery, 1909
    Species: *Eudistoma viride* Tokioka, 1955

Material Examined Colonies of this species were collected from small stones in North Break Water and Hare Island at a depth of two meters. Previously reported from Thoothukudi coast (Renganathan 1984a; Tamilselvi 2008; Tamilselvi et al. 2011).

Distribution Gulf of Mannar, Southeast coast, India.

World Distribution Central Indo-Pacific.

Description Colonial ascidian. Colonies are Greenish yellow in colour. Lobes of the colonies are closely packed. Colonies are free of epibionts. Black spots on either side of the oral siphon's basal region are the characteristic mark. No distinct constriction between thorax and abdomen (Fig. 9.20).

Remarks *E. viride* differs from *E. laysani* by the smaller size of its larvae and the dark pigment spots at the base of the oral siphon and it also differ from *E. muscosum* in which a dark spot is found in between the siphons.

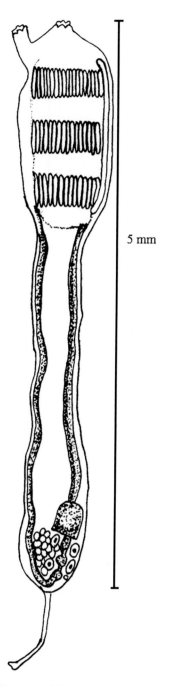

Fig. 9.19 Whole zooid of *Eudistoma tumidum*

Fig. 9.20 *Eudistoma viride*
—**a** Whole zooid with
abdomen, **b** Larva

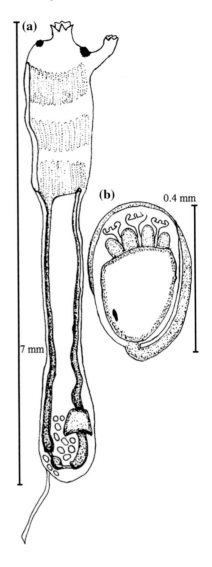

Synoicum citrum Kott, 1992 (Plate 9.4h)

Class: Ascidiacea
Suborder: Aplousobranchia
Family: Polyclinidae Milne-Edwards, 1841
Genus: *Synoicum* Phipps, 1774
Species: *Synoicum citrum* Kott, 1992

Material Examined A single colony of this species was found attached to large embedded rock and was collected from Tiruchendur at a depth of one meter.

Distribution New Report. Gulf of Mannar, Southeast coast, India.

World Distribution Australia.

Description The colony is rounded cushion, 2 cm in height and 3 cm in maximum dimension. The colony is attached to the substrate by a small part of the basal surface. The test is gelatinous and translucent. Sand sparsely distributed at the surface and crowded at the base. Zooids are crowded and open onto the upper surface. 5 mm long, the fleshy thorax and the abdomen together bring about half of the total length of the zooid. 10 rows of about 18–20 stigmata in the branchial sac. Atrial lip short and arises from the anterior rim of the atrial aperture. Gut loop is not twisted (Fig. 9.21).

Fig. 9.21 Whole zooid of *Synoicum citrum*

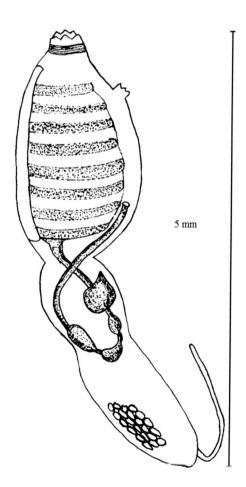

5 mm

Remarks This species is characterized by sessile, gelateneous colony as described by Kott (1992). In the present specimen sand is sparsely distributed at surface and crowded at the base but lacks in the test of *Synoicum citrum* Kott (1992).

Synoicum galei (Michaelsen, 1930) (Plate 9.5a)

    Class: Ascidiacea
    Suborder: Aplousobranchia
    Family: Polyclinidae Milne-Edwards, 1841
    Genus: *Synoicum* Phipps, 1774
    Species: *Synoicum galei* (Michaelsen, 1930)

Material Examined Colonies of this species were collected at Mandapam from pillars of jetty at a depth of 1–2 m.

Distribution New Record. Gulf of Mannar, Southeast coast, India.

World Distribution Australia.

Description Colonies are sand free, flat topped cushions up to 3.0 cm in maximum dimension and 0.5 cm thick. The test is soft and translucent internally, but the outer surface is tough. Common cloacal apertures are sessile. Zooids are arranged in a circular system. Central part of the base of the colony is attached to the substrate. Thorax long and narrow with 14 rows of up to 10 stigmata. The abdomen and posterior abdomen is not separated by constriction (Fig. 9.22).

Remarks The broad ridges and furrows on the surface are reminiscent of those in *Aplidium crateriferum* and *A. lobatum*. Zooids are more closely resemble those of *Synoicum papilliferum*.

Polyclinum fungosum Herdman, 1886 (Plate 9.5b)

    Class: Ascidiacea
    Suborder: Aplousobranchia
    Family: Polyclinidae Milne-Edwards, 1841
    Genus: *Polyclinum* Savigny, 1816
    Species: *Polyclinum fungosum* Herdman, 1886

Material Examined Reported at Mandapam, Thoothukudi Harbour, Inigo Nagar and Tiruchendur. Colony of this species was collected from hull of boat, pillars of jetty and small stones at a depth of 1–2 m.

Distribution Gulf of Mannar, Southeast coast, India.

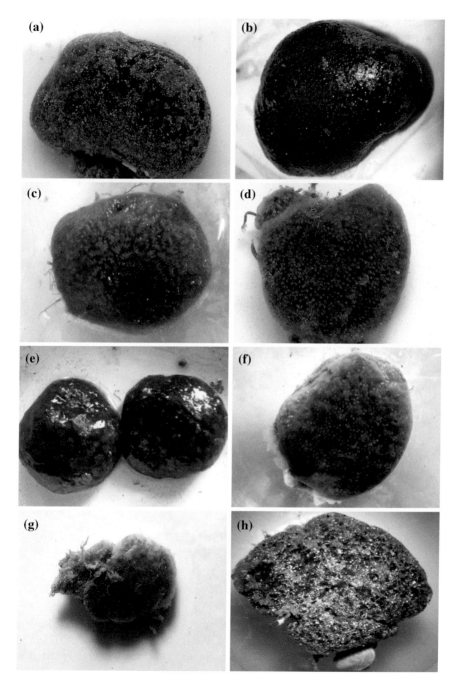

Plate 9.5 **a** *Synoicum galei* (ICM035), **b** *Polyclinum fungosum* (ICM036), **c** *Polyclinum glabrum* (ICM037), **d** *Polyclinum indicum* (ICM038), **e** *Polyclinum madrasensis* (ICM039), **f** *Polyclinum nudum* (ICM040), **g** *Polyclinum saturnium* (ICM041), **h** *Polyclinum solum* (ICM042)

Fig. 9.22 Whole zooid of
Synoicum galei

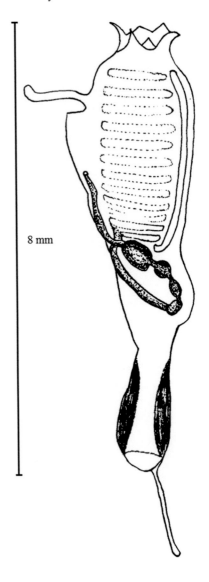

8 mm

World Distribution New South Wales (Lower East coast), Victoria (Bass Strait).

Description The Colour of the colony is sandy like brown or light black in living condition and dark brown in preservative. Colonies are cushions to about 7 cm in diameter and up to 1.5 cm thick. Test gelatinous, translucent internally. Sand is embedded in the surface test. The test is usually soft in preservative. Branchial lobes six, atrial lip is long originating from the body wall anterior to the atrial opening. There are 14 rows of up to 16 relatively short oval stigmata. No longitudinal folds in stomach (Fig. 9.23).

Fig. 9.23 Whole zooid of
Polyclinum fungosum

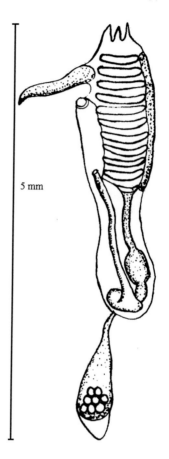

5 mm

Remarks In the present study *P. fungosum* Herdman, 1886 were first reported in Gopalapattinam coastal water in Palk Bay. Meenakshi (2004) reported in Gulf of Mannar. The characteristics of the present species colonies are cushions to about 7 cm in diameter and up to 1.5 cm thick. Test gelatinous, translucent internally. Sand is embedded in the surface test. Atrial lips long originated from the body wall anterior to the atrial opening. Stigmata are short.

Polyclinum glabrum Sluiter, 1895 (Plate 9.5c)

   Class: Ascidiacea
   Suborder: Aplousobranchia
   Family: Polyclinidae Milne-Edwards, 1841
   Genus: *Polyclinum* Savigny, 1816
   Species: *Polyclinum glabrum* Sluiter, 1895

Material Examined Reported at Mandapam, Thoothukudi Harbour and Inigo Nagar. Colony of this species was collected from pillars of jetty, hull of boat and small stones at a depth of 1–2 m. Previously reported from Thoothukudi coast (Meenakshi 2003).

Distribution Gulf of Mannar, Southeast coast, India.

World Distribution Indonesia, Northern Territory (North coast), Queensland (Great Barrier Reef, Northeast coast), Western Australia (Northwest coast).

Description The colony is dark black or dark brown in living condition and brown in preservative. No longitudinal folds in stomach, branchial lobes six, Ovary in post abdomen. Abdomen and post abdomen separated by constriction, Gut loop twisted. The test is usually soft in preservative. Colonies are soft, transparent test with entirely free of sands. Colonies are black in preservative. No sands embedded in the surface of the test. Zooids are with large thorax. A small horizontal gut loop and a spherical posterior abdomen. Atrial languet originates from body wall anterior to the aperture. Distinct brachial papillae are present. Atrial lip is long and moderately wide with fine longitudinal muscles. Thorax with 12 rows of up to 18 oval stigmata (Fig. 9.24).

Remarks The present colony collected from Gopalapattinam coastal waters in southeast coast of India. Colonies are soft, transparent test with entirely free of sands. Colonies are black in preservative. No sands embedded in the surface of the test. Zooids are with large thorax and small horizontal gut loop and a spherical posterior abdomen. Atrial languet originates from body wall anterior to the aperture. Distinct brachial papillae are present. The specimens resemble closely those previously assigned to this species by Sluiter (1895) and Kott (2002).

Polyclinum indicum Sebastian, 1954 (Plate 9.5d)

> Class: Ascidiacea
> Suborder: Aplousobranchia
> Family: Polyclinidae Milne-Edwards, 1841
> Genus: *Polyclinum* Savigny, 1816
> Species: *Polyclinum indicum* Sebastian, 1954

Material Examined Colonies of this species were collected from oyster cages in Thoothukudi harbour and from small stones in Inigo Nagar at a depth of 5–6 m and 1–2 m respectively. Previously reported from Madras coast (Sebastian 1954) and Thoothukudi coast (Tamilselvi 2008).

Distribution Gulf of Mannar, Southeast coast, India.

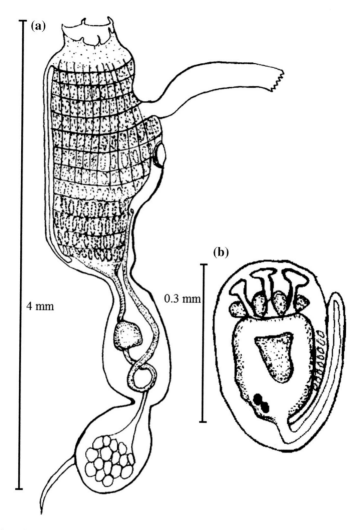

Fig. 9.24 *Polyclinum glabrum*—**a** Whole zooid with posterior abdomen, **b** Larva

Description The colony is greenish brown or brown in living condition. Colonies are larger, soft and mushroom shaped, attached by a small part of the base of the colony. Sand encrusts the sides and under surfaces an in patch on the upper surface and is sparse internally. Zooids are narrow. Thorax is small with horizontal gut loop. The branchial sac is narrow with 13 rows of 14 short oval stigmata. Branchial lobes six. Atrial languet originates from body wall anterior to the aperture. No longitudinal folds in stomach, Ovary in post abdomen, Abdomen and post abdomen separated by constriction, Gut loop twisted. (Figure 9.25).

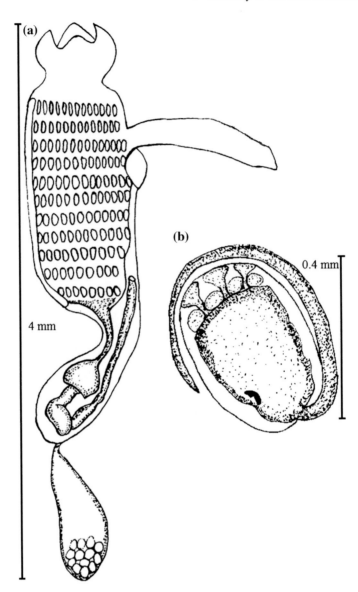

Fig. 9.25 *Polyclinum indicum*—**a** Whole zooid with posterior abdomen, **b** Larva

Remarks The present species agrees with the description of *P. indicum* Sebastian, 1954 in all respects. The characteristics of this species are no longitudinal folds in stomach. Twisted gut loop. Larger, soft and mushroom shaped colonies. This species differs from *P. madrasensis* in the presence of sand encrusts the sides and under surfaces an in patch on the upper surface and is sparse internally. Zooids are narrow. Thorax is small with horizontal gut loop. The branchial sac is narrow with 13 rows of 14 short oval stigmata.

Polyclinum madrasensis Sebastian, 1954 (Plate 9.5e)

Class: Ascidiacea
Suborder: Aplousobranchia
Family: Polyclinidae Milne-Edwards, 1841
Genus: *Polyclinum* Savigny, 1816
Species: *Polyclinum madrasensis* Sebastian, 1954

Material Examined Colony of this species was collected at Inigo Nagar from small stones at a depth of 1–2 m. Previously reported from Madras, Thoothukudi coast (Sebastian 1952; Tamilselvi 2008; Tamilselvi et al. 2011).

Distribution Gulf of Mannar, Southeast coast, India.

Description The colonies are hard cushions to 3 cm long, usually sand free, but sand particles embedded at the bottom. The colonies are white or yellowish white in living condition and dark brown in preservative. The test is usually soft in preservative. Colonies are cushions to about 6 cm in diameter and up to 1.5 cm thick. Test gelatinous, translucent internally. Colonies are black in preservative. No sands embedded in the surface of the test. Zooids are long. Atrial lip is long originated from the body wall anterior to the atrial opening. There are 12–14 rows of up to 14 relatively short oval stigmata. It differs from other genera of Polyclinidae by the following characters. No longitudinal folds in stomach, branchial lobes six, Ovary in post abdomen, Abdomen and post abdomen separated by constriction, Gut loop twisted (Fig. 9.26).

Remarks *P. madrasensis* reported on Madras coast by Sebastain, 1952. Native to India. This species is characterized by the presence of six branchial lobes. Test gelatinous, translucent internally. Colonies are black in preservative. No sands embedded in the surface of the test. Zooids are long with carried eggs. Atrial lip is long originated from the body wall anterior to the atrial opening. There are 12–14 rows of up to 14 relatively short oval stigmata.

Polyclinum nudum Kott, 1992 (Plate 9.5f)

Class: Ascidiacea
Suborder: Aplousobranchia
Family: Polyclinidae Milne-Edwards, 1841
Genus: *Polyclinum* Savigny, 1816
Species: *Polyclinum nudum* Kott, 1992

Fig. 9.26 Whole zooid of
Polyclinum madrasensis

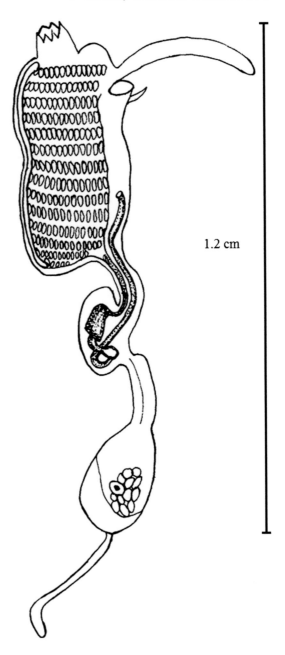

1.2 cm

Material Examined Colonies of this species were collected from pillars of jetty at a depth of 1–2 m in Mandapam. Previously reported from Thoothukudi coast (Meenakshi 1998b).

Distribution Gulf of Mannar, Southeast coast, India.

World Distribution New South Wales (Lower East coast).

Description Colony is cushion shaped dark black in preservative. The colony greenish brown or brown in living condition and brown in preservative. No sands on the either surface of the test or embedded within the colony. No longitudinal folds in stomach. Branchial lobes six. Abdomen and post abdomen separated by constriction. Twisted gut loop. Atrial languet originating from the upper rim of the atrial aperture. Long club shaped protruded from the surface on conical posterior abdomen is present. Ovary in post abdomen. Cloacal apertures are elevations (Fig. 9.27).

Remarks This species being reported for the first time in India. The present species agrees with the *P. nudum* Kott, 1992 in all respects. The characteristic features of the species are cushion shaped colonies and dark black in preservative. Cloacal apertures are protruded from the surface on conical elevations, no sands on either surface of the test or embedded within the colony. Long club shaped posterior abdomen is present. Cloacal apertures are protruded from the surface on conical elevations.

Polyclinum saturnium Savigny, 1816 (Plate 9.5g)

  Class: Ascidiacea
  Suborder: Aplousobranchia
  Family: Polyclinidae Milne-Edwards, 1841
  Genus: *Polyclinum* Savigny, 1816
  Species: *Polyclinum saturnium* Savigny, 1816

Material Examined Colonies of this species were collected at Mandapam and Colachel from hull of boat at a depth of 1–2 m. Previously reported from Thoothukudi coast (Meenakshi 1998b).

Distribution Gulf of Mannar, Southeast coast and Southwest coast, India.

World Distribution North Atlantic Ocean.

Description Colonies are cushions up to 2 cm in diameter, with sand throughout the surface. The internal test is soft and translucent. Light brown in preservative. Zooids arranged throughout the test in a circular system. Branchial siphon in upright, centered and with 10 lobes. Zooids are about 3 mm long with relatively long thorax and a long neck joining the posterior abdomen to the abdomen. Long atrial languet with 5–6 min pointed papillae. Atrial lip arising from the upper rim of

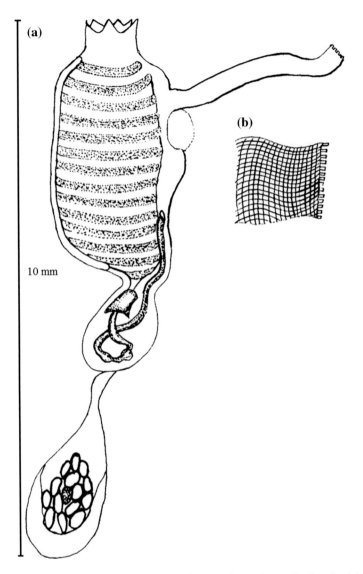

Fig. 9.27 *Polyclinum nudum*—**a** Whole zooid with posterior abdomen, **b** Tip of atrial languet showing mesh like structure

the atrial aperture. Zooids have 12 rows of up to 16 closely packed short oval stigmata. Well matured embryo in the peri-branchial cavity can be seen (Fig. 9.28).

Remarks The absence of brood pouch distinguishes these species from *Polyclinum tsutsuii* and *P. marsupiale*, *P. incrustum* and *P. solum* also sandy species but they have narrower branchial sac.

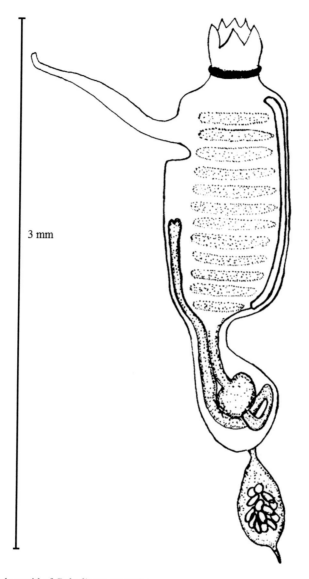

Fig. 9.28   Whole zooid of *Polyclinum saturnium*

Polyclinum solum Kott, 1992 (Plate 9.5h)

Class: Ascidiacea
Suborder: Aplousobranchia
Family: Polyclinidae Milne-Edwards, 1841
Genus: *Polyclinum* Savigny, 1816
Species: *Polyclinum solum* Kott, 1992

Material Examined Colony of this species was collected at Mandapam from hull of boat at a depth of 1–2 m. Previously reported by Meenakshi (2003).

Distribution Gulf of Mannar, Southeast coast, India.

Description The colonies are rounded cushions to 4 cm in greater extent and up to 1 cm height. A layer of sand is distributed in surface test. The internal test is soft, transparent and free of sand grains. Zooids are arranged in double rows surrounding the cloacal apertures. Zooids are slender in shape and 5–6 mm long. Thorax is long, about half of the length of zooid. The posterior abdomen is narrow and relatively long, more or less club-shaped. Branchial sac is wide with 14–16 rows of about 10–12 oval stigmata. Atrial tongue is long and narrow extending from the body wall (Fig. 9.29).

Remarks Characteristics of the present species are the single layer of sand externally, the conical protruberant cloacal apertures and absence of branchial papillae resembles the character of this species. *Polyclinum tenuatum* has similar but smaller crowded circular system to the present species, but its zooids are smaller, presence of branchial papillae and shorter thorax. *P. constellatum* has sand in the surface test but however it has very numerous stigmata and conspicuous branchial papillae present.

Polyclinum sundaicum (Sluiter, 1909) (Plate 9.6a)

Class: Ascidiacea
Suborder: Aplousobranchia
Family: Polyclinidae Milne-Edwards, 1841
Genus: *Polyclinum* Savigny, 1816
Species: *Polyclinum sundaicum* (Sluiter, 1909)

Material Examined Colony of this species was collected at Colachel from hull of boat at one meter depth.

Distribution New Report. Southwest coast, India.

World Distribution Central Indo-Pacific.

Description The colony is small, stalked, mushroom shaped with gelatinous and translucence sand free test containing one or more circular system. Large branchial papillae are present. The branchial sac is narrow with atrial opening. The pointed atrial lip of zooids is moderately narrow originated from the upper border of atrial opening. Abdomen is short. Limited number of male follicles surrounding the ovary is observed in posterior abdomen (Fig. 9.30).

Fig. 9.29 Whole zooid of
Polyclinum solum

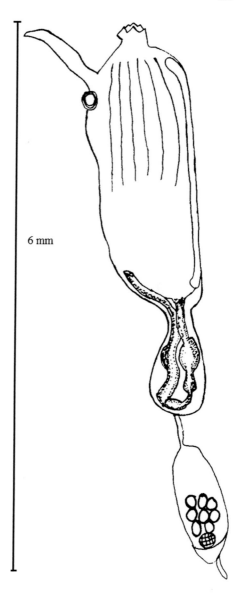

6 mm

Remarks These species distinguished from *Polyclinum fungosum* and *P. solum* (Kott 1992) by their large branchial papillae and stalked colonies. Few colonies of these species are collected from the Colachel station for the first time.

Polyclinum tenuatum Kott, 1992 (Plate 9.6b)

   Class: Ascidiacea
   Suborder: Aplousobranchia

Plate 9.6 **a** *Polyclinum sundaicum* (ICM043), **b** *Polyclinum tenuatum* (ICM044), **c** *Aplidiopsis confluata* (ICM045), **d** *Trididemnum caelatum* (ICM046), **e** *Trididemnum clinides* (ICM047), **f** *Trididemnum cyclops* (ICM048), **g** *Trididemnum savignii* (ICM049), **h** *Trididemnum vermiforme* (ICM050)

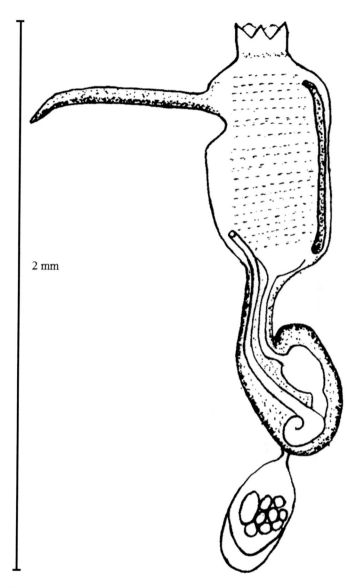

Fig. 9.30 Whole zooid of *Polyclinum sundaicum*

Family: Polyclinidae Milne-Edwards, 1841
Genus: *Polyclinum* Savigny, 1816
Species: *Polyclinum tenuatum* Kott, 1992

Material Examined Colonies of this species were collected at Mandapam and Tiruchendur from pillars of jetty and large embedded rock respectively at a depth of 1–2 m. Previously reported by Meenakshi (2003) and Abdul Jaffar Ali (2010).

Distribution Gulf of Mannar, Southeast coast, India.

World Distribution Australia.

Description The colonies are fleshy cushion sheets up to 6 cm in maximum extent with rounded border. The colonies are fixed to the substrate by the whole of the under surface. Dark green in colour. The test is gelatinous. The thorax and abdomen are together about 4 mm long. The posterior abdomen is short and sac like. Long atrial languet is protruded forwards from the upper rim of the atrial siphon. 5–6 minute pointed papillae form a fringe along the straight tip of the atrial lip. 13 rows of up to 12 relatively short oval stigmata with conspicuous conical branchial papilla. The gut loop is twisted and the distal part of the loop curves forward as is a characteristic for the genus (Fig. 9.31).

Remarks This species is first reported by Kott (1992) in Australian waters. The present observation has similarities with *P. solum* but the later has a long thorax with more rows of fewer stigmata and lacks branchial papilla. This species is distinguished from *P. tenuatum* n. sp. by the smaller zooids. In *P. tenuatum* n. sp. the similar branchial papilla and atrial lip are present but its zooids are larger.

Aplidiopsis confluata Kott, 1992 (Plate 9.6c)

> Class: Ascidiacea
> Suborder: Aplousobranchia
> Family: Polyclinidae Milne-Edwards, 1841
> Genus: *Aplidiopsis* Lahille, 1890
> Species: *Aplidiopsis confluata* Kott, 1992

Material Examined Colony of this species was collected at Mandapam from pillar of jetty at 1–2 m depth. Previously reported at Vizhinjam Bay (Abdul Jaffar Ali et al. 2010).

Distribution Gulf of Mannar, Southeast coast, India.

World Distribution Australia.

Description The colony is soft and cushion like about 3 cm in diameter attached by a thick, wrinkled stalk and some sand on it. The test is soft and translucent internally. Zooids are crossed one another in the center of the colony. Zooids are

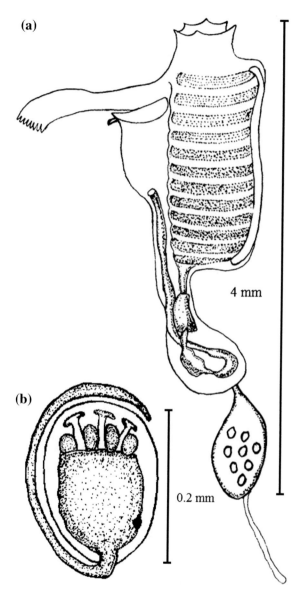

Fig. 9.31 *Polyclinum tenuatum*—**a** Whole zooid with posterior abdomen, **b** Larva

about 1 cm long with maximum length of 3 mm of thorax and the abdomen is about
1.5 mm. Posterior abdomen is wide and long crowded with gonads and pigments.
Branchial aperture has 6 small pointed lobes. There is a long, flat wide tongue from
the body near the anterior edge of the first row of stigmata. A sphincter muscle is

present at the atrial opening. There are 16 rows of up to 14 stigmata. The gut loop is vertical. The duodenum expands in diameter to its junction with short proximal part of the intestine. The posterior abdomen is long extending into the center of the colony. Gonads are observed at the middle part of the posterior abdomen. The gonads are of various sizes randomly distributed in the cluster and they are not in regular series (Fig. 9.32).

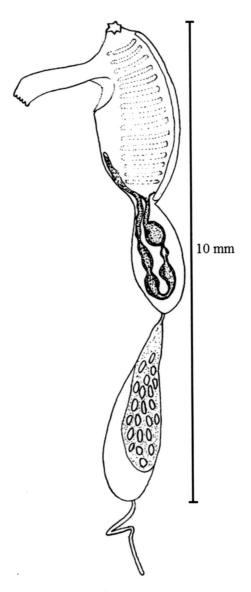

10 mm

Fig. 9.32 Whole zooid of *Aplidiopsis confluata*

Remarks The species is unique in its soft sand free test, crowded zooids and their long posterior abdomen each with long bunch of testis follicles. It resembles the other *Aplidiopsis mammillata* n. sp. in its long longitudinal muscles confined to the anterior part of the long delicate thorax and the absence of branchial papillae. This genus resembles those of *Polyclinum*. The zooids differ mainly in the absence of branchial papillae, the relatively large and vertical gut loop. The relatively large, vertical gutloop and the larger zooids of *Aplidiopsis* are very often, the only reliable characters to distinguish it from *Polyclinum*.

Trididemnum caelatum Kott, 2001 (Plate 9.6d)

Class: Ascidiacea
Suborder: Aplousobranchia
Family: Didemnidae Giard, 1872
Genus: *Trididemnum* Della Valle, 1881
Species: *Trididemnum caelatum* Kott, 2001

Material Examined This species was collected at Mandapam from pillar of jetty at 1–2 m depth.

Distribution Gulf of Mannar, Southeast coast, India.

World Distribution Australia.

Description Colony is thin, like encrusting sheet with a single layer of sand externally. Unusual distribution of spicules in the test can be seen. The surface test is soft owing to the absence of spicules also around the common cloacal aperture. More numerous spicules with pointed conical rays can be seen. Zooids are small with short branchial siphon, its rim divided into 6 triangular lobes. Atrial siphon is posteriorly directed. Larvae present in the basal test with long spherical larval trunk (Fig. 9.33).

Remarks Unusual arrangement of large, long, tapering spicules in the test, particularly around the common cloacal aperture is the characteristic feature of this species.

Trididemnum clinides Kott, 1977 (Plate 9.6e)

Class: Ascidiacea
Suborder: Aplousobranchia
Family: Didemnidae Giard, 1872
Genus: *Trididemnum* Della Valle, 1881
Species: *Trididemnum clinides* Kott, 1977

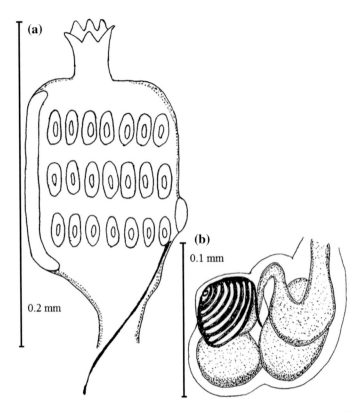

Fig. 9.33 *Trididemnum caelatum*—**a** Thorax showing the stigmata, **b** Abdomen showing the vas deferens

Material Examined Reported at Mandapam and Colachel. Colony of this species was collected from pillar of jetty and hull of boats at one meter depth. Previously reported by Meenakshi (2003).

Distribution Gulf of Mannar, Southeast coast, India.

World Distribution South Pacific Ocean.

Description The colony is thin encrusting sheet extending about 2–3 cm and robust. Live colonies are grayish green in colour whereas preserved is White. Test is soft. Stellate spicules are sparsely distributed throughout the test, having 12 pointed rays. Small cluster of spicules around the atrial opening is notable. Branchial siphon is upright centered with 6 lobes. Atrial siphon is short and wide at the bottom of the thorax. Zooids are short with 3 rows of about 6–7 stigmata. Vas deferens is coiled (Fig. 9.34).

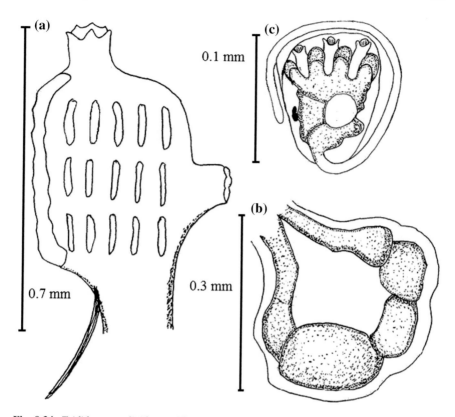

Fig. 9.34 *Trididemnum clinides*—**a** Thorax showing the retractor muscle, **b** Abdomen, **c** Larva

Remarks The colonies are distinctive with its soft test provided with embedded Prochloron, a chlorophyte symbiont can be seen throughout the test is the characteristic feature of this species. Spicules are evenly but sparsely distributed. Margin of some spicules has round tipped and others with pointed conical rays. Small group of spicules enclose the atrial opening into the cloacal cavity.

Trididemnum cyclops Michaelsen, 1921 (Plate 9.6f)

    Class: Ascidiacea
    Suborder: Aplousobranchia
    Family: Didemnidae Giard, 1872
    Genus: *Trididemnum* Della Valle, 1881
    Species: *Trididemnum cyclops* Michaelsen, 1921

Material Examined Colony of this species was collected at Mandapam from pillar of jetty at one meter depth. Previously reported by Meenakshi (2003).

Distribution Gulf of Mannar, Southeast coast, India.

World Distribution South Pacific Ocean.

Description Colony thin encrusting sheet, up to 1 cm long. A layer of bladder cells conspicuous around the outer margin of the colony. Spicules stellate, uniformly distributed and crowded with pointed rays. Orange coloured pigment cells throughout the test. Zooids short, 1–1.5 mm long, with 3 rows of up to 6–7 stigmata. Retractor muscle short. Branchial siphon is upright with 6 lobes (Fig. 9.35).

Remarks This species is distinguished by its small colonies with single system, endostylar pigment cap, 6 coils of the vas deferens, retractor from upper half of the

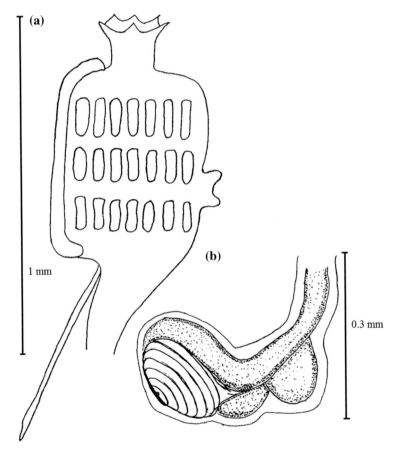

Fig. 9.35 *Trididemnum cyclops*—**a** Thorax showing the retractor muscle, **b** Abdomen showing the vas deferens

oesophageal neck and larvae with 2 adhesive organs and presence of prochloron. *Trididemnum paracyclops* is a related species with an endostylar pigment cap and 2 larval adhesive organs, but it has larger colonies with numerous systems. *T. symbioticum* having stellate spicules with conical to round tipped rays, an endostylar pigment cap. However *T. symbioticum* with posteriorly oriented atrial siphons, spicules crowded in the test with green plant cells, reddish pigment and embedded symbionts, is more like *T. clinides*, which lacks the endostylar pigment cap.

Trididemnum savignii (Herdman, 1886) (Plate 9.6g)

> Class: Ascidiacea
> Suborder: Aplousobranchia
> Family: Didemnidae Giard, 1872
> Genus: *Trididemnum* Della Valle, 1881
> Species: *Trididemnum savignii* (Herdman, 1886)

Material Examined Reported at Mandapam and Colachel. Colony of this species was collected from pillar of jetty and hull of boats at one meter depth. Previously reported by Meenakshi (2003).

Distribution Gulf of Mannar, Southeast coast, India.

World Distribution Indonesia, Bermuda, Florida, Northern Territory (North coast), Queensland (Northeast coast), Western Australia (Northwest coast); West Indies.

Description Large, smooth surface, fleshy colony is about 3 mm thick. Crowded bladder cells are found on the superficial layer of test. Pigmented cells are scattered among the bladder cells of the upper layer whereas the globular black vesicles are in the basal part of the test. Spicules are large, compact and distributed from the upper half and extend to the middle thorax of the zooid. Large zooids having large branchial siphon at terminal end with 6 pointed round lobes and short atrial siphon laterally. Presence of pigment cap is on the endostylar region (Fig. 9.36).

Remarks Presence of obvious superficial bladder cells layer and faecal pellets at the base of the colony resembles the colony described by Kott (New species of Didemnidae). *Trididemnum areolatum* has similar sharply pointed spicules as *T. savignii*. *T. savignii* has thick surface bladder layer, but it distinguished from that species by its sharply pointed spicules rays and by spicule distribution. *T. natalense* has a layer of spicules beneath the cloacal canals and this together with its smaller spicules distinguishes it. *T. pigmentatum* has spicule like those of present species, with similar strong conical pointed rays.

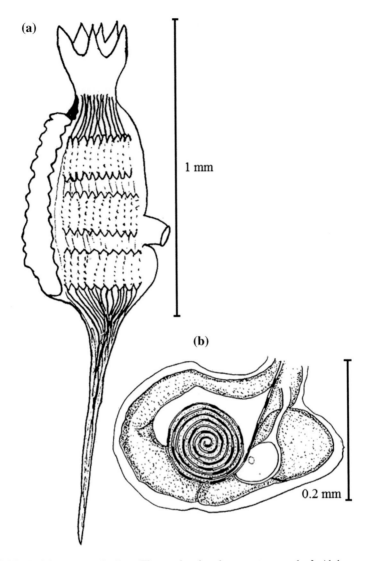

Fig. 9.36 *Trididemnum savignii*—**a** Thorax showing the retractor muscle, **b** Abdomen showing the vas deferens

Trididemnum vermiforme Kott, 2001 (Plate 9.6h)

Class: Ascidiacea
Suborder: Aplousobranchia
Family: Didemnidae Giard, 1872
Genus: *Trididemnum* Della Valle, 1881
Species: *Trididemnum vermiforme* Kott, 2001

Material Examined Colony of this species was collected at Mandapam from pillar of jetty at one meter depth.

Distribution Gulf of Mannar, Southeast coast, India.

World Distribution Australia.

Description Colony is thick and fleshy, up to 5 to 7 cm in maximum dimension. Conspicuous circular common cloacal apertures along the surface of the test. The surface test is folded forming the shape of lobes. Spicules are large and stellate with conical pointed rays. Zooids are small, up to 1 mm long with relatively short branchial siphons. Orange-coloured embryos are crowded in the surface layer of the test (Fig. 9.37).

Remarks This species differs from *T. savignii* by its stellate spicules with 7–9 conical pointed rays. Besides spicules are large but smaller than the latter one. *T. cerebriforme* also resembles the present species but is distinguished by its restricted spicule distribution and 4 larval ectodermal ampullae. *T. nobile* is also similar in many aspects but has thicker colony lobes, smaller spicules.

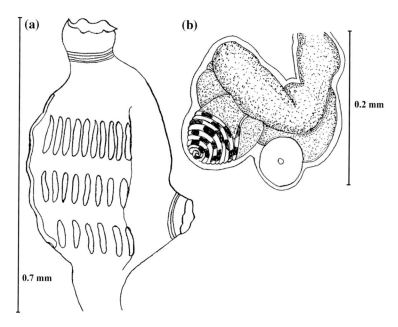

Fig. 9.37 *Trididemnum vermiforme*—**a** Thorax showing the stigmata, **b** Abdomen showing the vas deferens

Didemnum psammathodes (Sluiter, 1895) (Plate 9.7a)

Class: Ascidiacea
Suborder: Aplousobranchia
Family: Didemnidae Giard, 1872
Genus: *Didemnum* Savigny, 1816
Species: *Didemnum psammathodes* (Sluiter, 1895)

Material Examined Reported at Mandapam, Muthunagar beach, Inigo Nagar, Thoothukudi Harbour, Hare Island, North Break Water, Veerapandianpattinam, Tiruchendur, Chinna Muttom and Colachel. Several large sheet or mat like colonies of this species were observed in small stones, embedded rocks and pillars of jetty at shallow water regions. This species was also previously reported at Thoothukudi coast, Kanyakumari, Rameshwaram, Kurusadai Island, Vizhinjam (Renganathan 1981a; Meenakshi 1997; Tamilselvi 2008; Abdul Jaffar Ali et al. 2009; Tamilselvi et al. 2011, 2013).

Distribution Gulf of Mannar, Southeast coast, India.

World Distribution Japan, Malaysia, Indonesia, New Zealand, New South Wales (Central East coast), Queensland (Central East Coast, Great Barrier Reef, Northeast Coast), Victoria (Bass Strait); West Indian Ocean, Red Sea.

Description This species is abundant in all the habitats. Colony forms thin encrusting sheets spreading over the substrates with characteristically restricted thoracic common cloacal center. The colony is brown or light yellowish brown in living condition. Faecal pellets are embedded throughout the colony. Spicules occur throughout the surface test and around the branchial apertures, but not crowded. Zooids are very small, less than 1 mm long with 4 rows of stigmata. Atrial opening is wide (Fig. 9.38).

Remarks *D. psammathodes*, (Sluiter 1895) were first reported in Gopalapattinam coastal water in Palk Bay. Previous data report reflecting nobody surveys this area. The material examined confirms well to the earlier description Renganathan (1981), except for minor variation in the relative measure of the colonial length and larvae size. The total length of colony varied between 10–25 cm and size of larvae 0.5 mm. The colony length is 4–17 cm and larvae size 0.5 mm but not clear. Accumulation of faecal pellets throughout in the colony is the characteristic feature.

Didemnum spadix Kott, 2001 (Plate 9.7b)

Class: Ascidiacea
Suborder: Aplousobranchia
Family: Didemnidae Giard, 1872

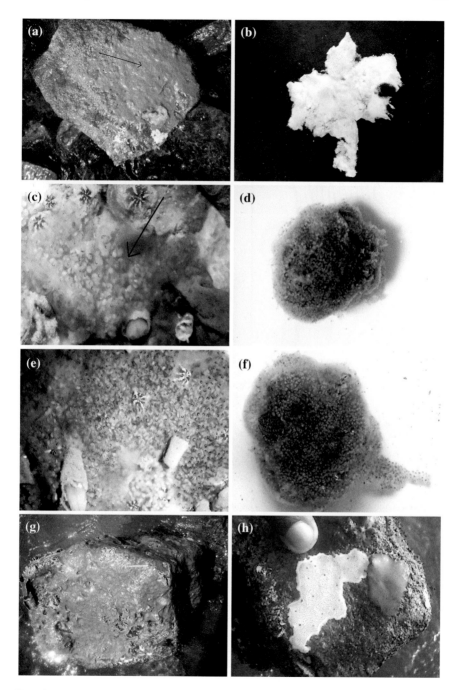

Plate 9.7 a *Didemnum psammathodes* (ICM051), **b** *Didemnum spadix* (ICM052), **c** *Diplosoma listerianum* (ICM053), **d** *Diplosoma macdonaldi* (ICM054), **e** *Diplosoma simileguwa* (ICM055), **f** *Diplosoma swamiensis* (ICM056), **g** *Lissoclinum bistratum* (ICM057), **h** *Lissoclinum fragile* (ICM058)

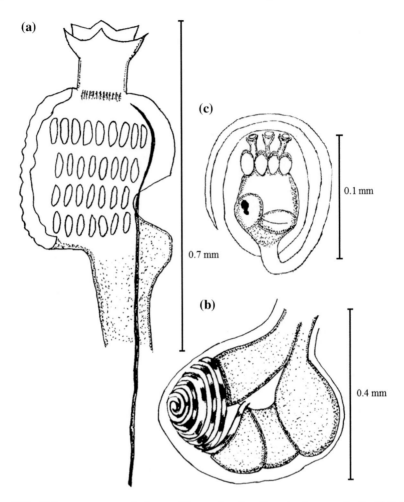

Fig. 9.38 *Didemnum psammathodes*—**a** Thorax showing the retractor muscle, **b** Abdomen showing the vas deferens, (**c**) Larva

Genus: *Didemnum* Savigny, 1816
Species: *Didemnum spadix* Kott, 2001

Material Examined Colony of this species was collected at Mandapam from pillar of jetty at one meter depth.

Distribution Gulf of Mannar, Southeast coast, India.

World Distribution Australia.

Description The colony is white, flat and the size is about 3–4 cm. Brown coloured pigment cells are present in the test. Spicules are crowded throughout the colony. Stellate spicules with 10 large pointed rays. The branchial siphon is upright centered with 6 branchial lobes. Zooids are short with linear with 4 rows of stigmata and 7 stigmata in a row (Fig. 9.39).

Remarks Large vesicular cells regularly arranged around siphonal openings and large undivided testis with 8 coils of the vas deferens around it are the distinctive features of this species as reported by Kott (2001) Spicules of the present species have about the same number of rays as *Didemnum fuscum*, but are shorter and more conical (like the rays of *D. sordium*). *D. sordium* and *D. fuscum* lack the large vesicular cells in the surface. *D. spadix* resembles the most character of *D. sordium* including small zooid with few stigmata and numerous coil of vas deferens.

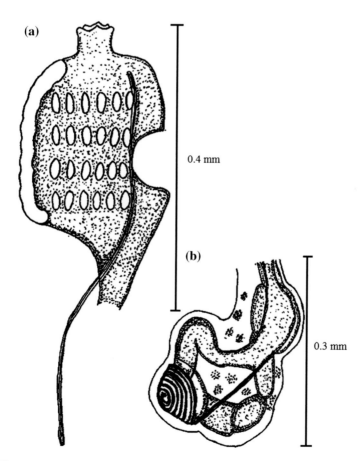

Fig. 9.39 *Didemnum spadix*—**a** Thorax showing the retractor muscle, **b** Abdomen showing the vas deferens

Diplosoma listerianum (Milne-Edwards, 1841) (Plate 9.7c)

    Class: Ascidiacea
    Suborder: Aplousobranchia
    Family: Didemnidae Giard, 1872
    Genus: *Diplosoma* Macdonald, 1859
    Species: *Diplosoma listerianum* (Milne-Edwards, 1841)

Material Examined Colony of this species was collected at Colachel from hull of boat fouling the other ascidian species at a depth of one meter. Previously reported from Gulf of Mannar, Southeast coast, India (Meenakshi 2003).

Distribution Southwest coast, India.

World Distribution South Africa, North Sea, France, Gulf of Mexico, Madagascar, Netherlands, North Atlantic Ocean, North Pacific Ocean, South Atlantic Ocean, Tanzania, Central Atlantic Ocean, New Zealand.

Description The colony is very thin encrusting on substrate as transparent sheet. The test is delicate, glassy transparent and colourless in nature. Zooids crowded in the test. Zooids are small up to 0.8–1.2 mm long, white with brownish pigments. Large leaf like branchial lobes that alternate with smaller intermediate lobe. Dark pigment found above the endostyle. Branchial sac with 4 rows of stigmata and 8–10 stigmata per row. Limited cloacal system. Light yellow coloured symbiotic cells (prochlorons) distributed throughout the test. Larva 0.8 mm long with three adhesive papillae. The stomach is globules and smooth-walled, leads to an intestine of large diameter. Obvious stolonic vessels extend from the ventral side of the abdomen. The post-pyloric part of the gut is bent ventrally. Two oval testis follicles lie against the dorsal side of the gut loop. The vas deferens is hooked around between them, extending anteriorly around the ascending limb of the gut loop. The ovaries and eggs are found behind the intestinal loop (Van Name 1945; Kott 2001). The larvae have a trunk 0.4–6 mm long, with three fixatory papillae (Hayward and Ryland, 1990) (Fig. 9.40).

Remarks Zooids appear as white dots. Presence of very delicate muscular process differs from *D. macdonaldi* and absence of prochloron from *D. simile*. The larvae of this specimen resemble the larvae of (Hayward and Ryland 1990) in the presence of three papillae. The larvae are about 0.35 to 6 mm long.

Diplosoma macdonaldi Herdman, 1886 (Plate 9.7d)

    Class: Ascidiacea
    Suborder: Aplousobranchia
    Family: Didemnidae Giard, 1872

Fig. 9.40 Whole zooid of
Diplosoma listerianum

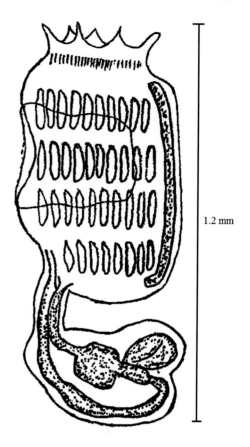

1.2 mm

Genus: *Diplosoma* Macdonald, 1859
Species: *Diplosoma macdonaldi* Herdman, 1886

Material Examined Colonies of this species were collected at Inigo Nagar and
Tiruchendur from small stones and a large embedded rock respectively at a depth of
1–2 m. Previously reported from Thoothukudi coast (Meenakshi 2003).

Distribution Gulf of Mannar, Southeast coast, India.

World Distribution Central Atlantic Ocean.

Description Colony is thin, delicate, encrusting on substrate even on accumulated
sand. The test is colourless and opaque in nature. Zooids are grey coloured. Large
thorax with wide atrial aperture.

Remarks Most of the characters of this species agree with *D. macdonaldi*. But the distinctive feature of this species is grey coloured smaller zooids embedded in the thin, soft, delicate test.

Diplosoma simileguwa Oka, Suetsuga & Hirose, 2005 (Plate 9.7e)

Class: Ascidiacea
Suborder: Aplousobranchia
Family: Didemnidae Giard, 1872
Genus: *Diplosoma* Macdonald, 1859
Species: *Diplosoma simileguwa* Oka, Suetsuga and Hirose, 2005

Material Examined Colony of this species was collected at Colachel from hull of boat fouling the other ascidian species at a depth of one meter.

Distribution New Report. Southwest coast, India.

World Distribution North Pacific Ocean.

Description The colony forms an irregular sheet, faded green and 3–4 cm long and 1–1.5 mm thick. Zooids are sparsely distributed throughout the colony in standing vertically. In matured colonies embryos are often seen through the transparent test. Prochloron cells are distributed on peribranchial area of the test. Zooids are about 1–1.5 mm long in preservative. Thorax is whitish whereas in yellowish colour. Each branchial siphon is 6 lobed and has 6 tentacles. Atrial apertures are comparatively wide. There are 4 rows of stigmata. The total number of stigmata per half branchial sac is 23 (6-5-6-6) arrangement. Dark brown colour pigments present in the branchial neck of many of the zooids. Testis and oocytes were found in some colonies (Fig. 9.41).

Remarks The present species bearing prochlorons as in the cases *of Diplosoma ooru, D. similis, D. multiplicatum* and *D. matie*. This species distinguishes from *D. variostigmatum* in the number of stigmata but resembles in the presence of prochlorons. Unique and constant number of stigmata of this species resembles the *D. ooru* (Hirose et al. 2005). Stigmata number where 5-6-5-4 in *D. ooru*, 6-6-6-5 in *D. similis* and *D. virens*.

Diplosoma swamiensis Renganathan, 1986 (Plate 9.7f)

Class: Ascidiacea
Suborder: Aplousobranchia
Family: Didemnidae Giard, 1872
Genus: *Diplosoma* Macdonald, 1859
Species: *Diplosoma swamiensis* Renganathan, 1986

Fig. 9.41 Whole zooid of
Diplosoma simileguwa

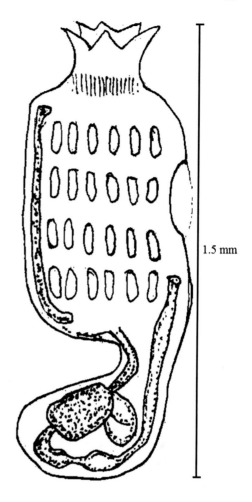

1.5 mm

Material Examined Colonies of this species were collected at Inigo Nagar from small stones at a depth of 1–2 m. Previously reported from Thoothukudi coast (Renganathan 1986b).

Distribution Gulf of Mannar, Southeast coast, India.

World Distribution West Indies, West Pacific Ocean.

Description The colony forms an irregular sheet, grey in colour and 6–8 cm in dia and 1–1.5 mm thick. Zooids are sparsely distributed throughout the colony in standing vertically. In matured colonies embryos are often seen through the transparent test. Prochloron cells are absent. Thorax is whitish whereas abdomen is yellowish in colour. Each branchial siphon is 6 lobed and has 6 tentacles. Atrial apertures are comparatively wide. There are 4 rows of stigmata. The total number of stigmata per half branchial sac is 20 (5-5-5-5) arrangement. Dark brown colour pigments present throughout the zooids. Testis and oocytes were found in some colonies.

Remarks The genus *Diplosoma* Macdonald, 1859, absence of spicules in the test and no coiled vas deferens. Test is gelatinous but firm contains colony is translucent. Presence of Prochloron is common. *D. swamiensis* is distinguished by a number of characters including its remarkable larvae with tubular adhesive organs (Kott 1990).

Lissoclinum bistratum (Sluiter, 1905) (Plate 9.7g)

    Class: Ascidiacea
    Suborder: Aplousobranchia
    Family: Didemnidae Giard, 1872
    Genus: *Lissoclinum* Verrill, 1871
    Species: *Lissoclinum bistratum* (Sluiter, 1905)

Material Examined Several large colonies of this species were observed from small stones in North Break Water at a depth of one meter. Previously reported by Meenakshi (2003).

Distribution Gulf of Mannar, Southeast coast, India.

World Distribution Djibouti, South Pacific Ocean, Tanzania.

Description Colonies are large extensive sheets up to 30 cm in maximum dimension and 5 mm in thickness. One or more common cloacal apertures on the surface. Spicules are crowded at the base and around the margins of the colony. A layer of carotenoid pigments in the upper layer of the surface test. Colonies are pinkish orange coloured in living condition and pale orange in preservative. Zooids are short, up to 1 mm long. A narrow retractor muscle projects from the posterior end of the thorax. Branchial sac has 4 rows of stigmata with 6 stigmata per row. Larvae in the basal test of the colony (Fig. 9.42).

Remarks Spicules are crowded at the base and around the margin of the colony. Symbiotic prochloron is crowded above the cloacal chamber. Spicules are spherical with flat ended cylindrical rays are the distinct features.

Lissoclinum fragile (Van Name, 1902) (Plate 9.7h)

    Class: Ascidiacea
    Suborder: Aplousobranchia

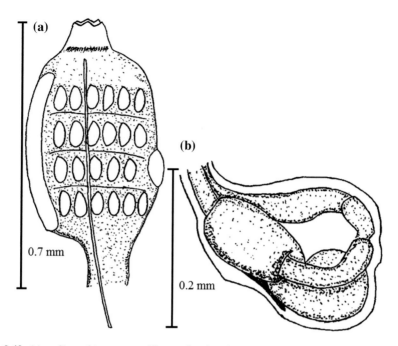

Fig. 9.42 *Lissoclinum bistratum*—**a** Thorax showing the retractor muscle, **b** Abdomen

Family: Didemnidae Giard, 1872
Genus: *Lissoclinum* Verrill, 1871
Species: *Lissoclinum fragile* (Van Name, 1902)

Material Examined Reported at Mandapam, Muthunagar beach, Inigo Nagar, Thoothukudi Harbour, Hare Island, North Break Water, Tiruchendur, Chinna Muttom and Colachel. Several colonies of this species were observed in small stones, embedded rocks and pillars of jetty at shallow water regions. Previously reported from Thoothukudi coast, Mandapam, Rameshwaram, Kurusadai Island, Singhale Island, Valinokkam and Ervadi, Bombay, Goa and Vizhinjam (Renganathan 1982a; Meenakshi 1997; Tamilselvi 2008; Tamilselvi et al. 2011).

Distribution Gulf of Mannar, Southeast coast, India.

World Distribution East Pacific Ocean of South America to California, West Indies, North Atlantic Ocean to Arctic Ocean and Northwest of Iceland, West Pacific Ocean.

Description The colony is thin, soft, encrusting with extensive cloacal cavity. The thorax is almost free, and the abdomen is included in the tunic. The colony is in varying shades of white colour. Larvae located in the basal lamina. The body wall is almost missing, visible at the level of a short atrial languet and posteriorly where it holds thoracic lateral organs at the basis of the fourth stigmata row. There is no fixative appendage. The testis has a straight sperm duct and ovary is present above the testis. The tadpole has 3 suckers and 4 small papillae.

Remarks *Lissoclinum fragile* is characterized by the whitish colour of the colony, even in preserved specimens, and no zooid pigmentation (Rodrigues et al. 1998). Also, the zooid is 1.5 mm long and the colony has low spicule density (Monniot 1983). Identification is easily confirmed by the bright yellow coloration of the abdominal region of individual zooids (Goodbody 2003). On the other hand, the spicules of the present species are very similar to the spicules of *L. perforatum*.

9.6 Discussion

In the present survey, 58 species from various families of class Ascidiacea from 11 stations were recorded (Fig. 9.43). The inventory of ascidian fauna is represented in Table 9.2. More number of species were recorded from Didemnidae (13), Polycitoridae (13), Polyclinidae (12) followed by Styelidae (6), Pyuridae (5), Ascidiidae (4), Perophoridae (3), Rhodosomatidae (1) and Molgulidae (1) from a variety of habitats. This record forms the inventory of ascidians from southern Indian coast.This record forms the inventory of ascidians from southern Indian coast. Many ascidian species had not been recorded from the area and resulted in several first records of ascidians from each station. All the species from Colachel, Chinna Muttom and Tiruchendur stations are recorded for the first time. Many species from Mandapam station also recorded for first time.

9.6.1 Didemnidae

Didemnidae is an abundant family in the most marine environments throughout the world. Indian costal water is no exception. Members of the family Didemnidae was distributed commonly in variety of substrata from different stations. Remarkable distribution of the ascidians species from this family was *Didemnum psammathodes*, and *Lissoclinum fragile* and was found to be occurred in maximum of 10 and 9 stations respectively and can be regarded as key species in the sessile communities. *D. psammathodes* is one of the most common and abundant Didemnids in both south west and south east coast of India. It is recorded in almost all kinds of habitats and substrata. It forms large sheet like colonies, muddy brown in colour. The colour is due to accumulation of faecal pellets in the test. *Lissoclinum fragile* is another key species forming thin mat colonies in many parts of the habitats. It is particularly abundant in Thoothukudi coastal waters. *Trididemnum savignii* and

Table 9.2 Inventory of ascidian fauna along the Southern Indian coastline

Family	Species	MPM	MB	IN	HI	NBW	TH	KP	VP	TR	CM	CL	References
Perophoridae	*Ecteinascidia garstangi*										x		(Renganathan 1984a)
	Ecteinascidia venui	x											(Meenakshi 1997; Tamilselvi et al. 2012)
	Perophora multiclathrata			x							x		(Renganathan 1983a; Tamilselvi et al. 2012)
Rhodosomatidae	*Rhodosoma turcicum*						x						(Meenakshi 1997)
Ascidiidae	*Phallusia arabica*						x						(Abdul Jaffar Ali et al. 2014; Meenakshi 2003; Tamilselvi et al. 2012)
	Phallusia nigra						x						(Abdul Jaffar Ali and Sivakumar 2007; Abdul Jaffar Ali et al. 2014; Meenakshi 1998; Tamilselvi et al. 2012)
	Ascidia gemmata						x						(Abdul Jaffar Ali et al. 2014; Krishnan et al. 1989; Tamilselvi et al. 2012)
	Ascidia sydneiensis						x						(Abdul Jaffar Ali et al. 2014; Meenakshi 1998; Tamilselvi et al. 2012)
Styelidae	*Botryllus purpureus*											x	(Meenakshi and Senthamarai 2006)
	Botryllus schlosseri											x	(Meenakshi and Senthamarai 2006)
	Botrylloides chevalense			x									(Renganathan 1984; Tamilselvi et al. 2012)
	Botrylloides magnicoecum											x	(Renganathan 1985; Tamilselvi et al. 2012)

(continued)

Table 9.2 (continued)

Family	Species	MPM	MB	IN	HI	NBW	TH	KP	VP	TR	CM	CL	References
	Symplegma oceania	x					x						(Abdul Jaffar Ali and Sivakumar 2007; Meenakshi 2003; Tamilselvi et al. 2012)
	Styela canopus	x	x	x			x			x	x	x	(Abdul Jaffar Ali and Sivakumar 2007; Abdul Jaffar Ali et al. 2010, 2014; Renganathan 1986)
Pyuridae	*Microcosmus curvus*		x								x		(Renganathan 1983; Tamilselvi et al. 2012)
	Microcosmus exasperatus	x					x						(Abdul Jaffar Ali and Sivakumar 2007; Abdul Jaffar Ali et al. 2010, 2014; Krishnan et al. 1989)
	Microcosmus propinquus	x					x						(Abdul Jaffar Ali and Sivakumar 2007; Meenakshi 2003)
	Microcosmus squamiger						x						(Abdul Jaffar Ali and Sivakumar 2007; Abdul Jaffar Ali et al. 2010, 2014; Meenakshi and Senthamarai 2007a)
	Herdmania momus						x	x	x				(Abdul Jaffar Ali and Sivakumar 2007; Abdul Jaffar Ali et al. 2010, 2014; Das 1936; Tamilselvi et al. 2012)
Molgulidae	*Molgula ficus*									x			(Meenakshi 2003)
Polycitoridae	*Eudistoma amplum*									x			(Meenakshi 2003)
	Eudistoma carnosum	x											Present study
	Eudistoma constrictum									x			(Meenakshi 2002)

(continued)

Table 9.2 (continued)

Family	Species	MPM	MB	IN	HI	NBW	TH	KP	VP	TR	CM	CL	References
	Eudistoma gilboviride	x	x	x	x	x	x			x			(Meenakshi 2003)
	Eudistoma lakshmiani		x	x									(Renganathan 1986; Tamilselvi et al. 2011)
	Eudistoma laysani			x									(Abdul Jaffar Ali and Sivakumar 2007; Abdul Jaffar Ali et al. 2010, 2014; Meenakshi 2002)
	Eudistoma microlarvum									x			(Meenakshi 2003)
	Eudistoma muscosum				x	x							(Meenakshi 2003)
	Eudistoma ovatum	x											(Meenakshi 2002)
	Eudistoma pyriforme	x											(Meenakshi 2003)
	Eudistoma reginum	x											(Meenakshi 2003)
	Eudistoma tumidum	x											(Meenakshi 2003)
	Eudistoma viride				x	x							(Abdul Jaffar Ali et al. 2014; Renganathan 1984; Tamilselvi et al. 2012)
Polyclinidae	*Synoicum citrum*	x								x			Present study
	Synoicum galei			x									Present study
	Polyclinum fungosum	x		x			x			x			(Abdul Jaffar Ali and Sivakumar 2007; Meenakshi 1998a)

(continued)

Table 9.2 (continued)

Family	Species	MPM	MB	IN	HI	NBW	TH	KP	VP	TR	CM	CL	References
	Polyclinum glabrum	x		x			x						(Meenakshi 2003)
	Polyclinum indicum			x			x						(Abdul Jaffar Ali and Sivakumar 2007; Sebastian 1954; Tamilselvi et al. 2012)
	Polyclinum madrasensis			x									(Abdul Jaffar Ali and Sivakumar 2007; Sebastian 1952; Tamilselvi et al. 2012)
	Polyclinum nudum	x											(Meenakshi 1998a)
	Polyclinum saturnium	x										x	(Abdul Jaffar Ali and Sivakumar 2007; Meenakshi 1998a)
	Polyclinum solum	x											(Meenakshi 2003)
	Polyclinum sundaicum											x	Present study
	Polyclinum tennuatum	x								x			(Abdul Jaffar Ali et al. 2010; Meenakshi 2003)
	Aplidiopsis confluata	x											(Abdul Jaffar Ali et al. 2010)
Didemnidae	*Trididemnum caelatum*	x											Present study
	Trididemnum clinides	x										x	(Abdul Jaffar Ali and Sivakumar 2007; Abdul Jaffar Ali et al. 2010; Meenakshi 2003: 89. Tamilselvi et al. 2012)
	Trididemnum cyclops	x											(Meenakshi 2003)

(continued)

Table 9.2 (continued)

Family	Species	MPM	MB	IN	HI	NBW	TH	KP	VP	TR	CM	CL	References
	Trididemnum savignii	x										x	(Meenakshi 2003)
	Trididemnum vermiforme	x											Present study
	Didemnum psammathodes	x	x	x	x	x	x		x	x	x	x	(Abdul Jaffar Ali et al. 2010, 2014; Renganathan 1981; Tamilselvi et al. 2012)
	Didemnum spadix	x											Present study
	Diplosoma listerianum											x	(Meenakshi 2003)
	Diplosoma macdonaldi			x						x			(Meenakshi 2003)
	Diplosoma simileguwa											x	Present study
	Diplosoma swamiensis			x									(Renganathan 1986; Tamilselvi et al. 2012)
	Lissoclinum bistratum					x							(Meenakshi 2003)
	Lissoclinum fragile	x	x	x	x	x	x			x	x	x	(Abdul Jaffar Ali et al. 2014; Renganathan 1982; Tamilselvi et al. 2012)

Note

MPM Mandapam; *MB* Muthunagar beach; *IN* Inigo Nagar; *HI* Hare Island; *NBW* North Break Water; *TH* Thoothukudi Harbour; *KP* Kayalpattinam; *VP* Veerapandianpattinam; *TR* Tiruchendur; *CM* Chinna Muttom; *CL* Colachel

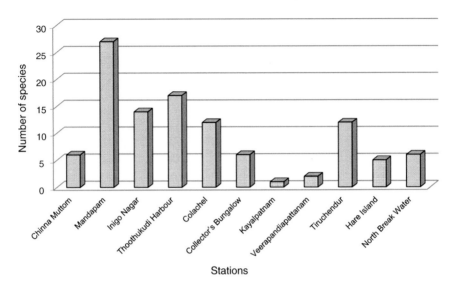

Fig. 9.43 Station-wise representation of ascidians along the Southern Indian coastline

T. clinides are found abundant and fouling the entire hull of boat particularly in Colachel station.

9.6.2   Styelidae

This is another large family which contains three types of ascidians such as colonial, synascidians and solitary, some of which are abundant in Colachel station. All the *Botryllus* and *Botrylloides* species recorded in the present study fouled the entire hull of boat. Solitary ascidian, *Styela canopus* was found to occur in most of the stations whereas, colonial ascidian, *Symplegma oceania* was observed in limited stations.

9.6.3   Polycitoridae

This important family is represented by 13 species. The colonial ascidian, *Eudistoma gilboviride* is wide spread throughout the south east coast but not conspicuous. But *E. viride* and *E. muscosum* were found abundant exclusively in Hare Island and North Break Water region. All other species of *Eudistoma* are abundant and found throughout the year in Mandapam water.

9.6.4 Pyuridae

About 5 species of ascidians were recorded from this family. Large number of solitary ascidian, *Microcosmus exasperatus* is orange and/or red in colour and is common in Thoothukudi harbour particularly in pearl oyster cages. Next to *M. exasperatus*, *M. squamiger* and *M. propinquus* were found abundant in the same location. Only one species from the genus *Herdmania* was found to distribute in many stations. Previous reports indicated that the simple ascidian *H. momus* was found as highly dominating species in Thoothukudi harbour area; in contrary, this species was rarely noted in the present study. In the present findings, even though this species was recorded from Kayalpattinam and Veerapandianpattinam it was not recorded for the past one decade.

9.6.5 Ascidiidae

Members of this large and important family are usually solitary. Out of four species recorded, *Phallusia nigra* was found abundant and throughout the year in Thoothukudi harbour. *P. arabica* and other members of the Ascidiidae are restricted to Thoothukudi harbour.

9.6.6 Polyclinidae

Large number of species (12) belong to this family has been observed next to Didemnidae and Polycitoridae. Among the 12 species recorded from this family, only two colonies of *Synoicum citrum* were collected from Tiruchendur station whereas few colonies of *S. galei* were observed at Mandapam station. The colonies were sand free, round cushion shaped and dark green in colour. Out of three species *P. fungosum* has wide distribution along the study areas.

9.6.7 Perophoridae

Only three species such as *Ecteinascidia garstangi*, *E. venui* and *Perophora multiclathrata* have been reported. Of these, *E. venui* was reported for the first time at Mandapam. Huge colonies with many bunches of this grape ascidian was found to harbour the pillers of Jetty. *E. garstangi* also was reported for the first time at Chinna Muttom station.

9.6.8   Rhodostomatidae

Only one species, *Rhodosoma turcicum* was found to foul oyster cages installed in Thoothukudi harbour area.

9.6.9   Molgulidae

Only one species, *Molgula ficus* was found to occur in the embedded rock at Tiruchendur station for the first time.

As ascidians leads a sedentary mode of life in adult, substratum is a significant ecological factor for them which influencing the survival and distribution in its marine environment. Number of scientists explored the ascidian fauna from different habitats of the marine ecosystem such as soft bottom community (Sanders 1960), soft muddy bottom (Bouchet 1962), rocky areas (Dybern 1963), artificial substrates such as hulls of ship, pearl oyster cages etc., (Abdul Jaffar Ali et al. 2009), muddy area, seagrass and coral bed (Tamilselvi 2008). Much attention has been directed to the dispersal capabilities and substrate selection of the short lived non-feeding larvae and are discussed by Bingham and Young (1991), Goodbody (1974), Marshall and Keough (2003).

The members of Didemnidae and Polycitoridae were commonly found in most of the stations in the present survey. *Didemnum* and *Eudistoma* are diverse and cosmopolitan and their distribution and dominance in various parts of the countries show wide-ranging adaptability to hydrographical parameters. Ascidians are one among, is a key ecological groups as the environmental variables influencing their recruitment, dispersal, survival and reproduction. Many workers reported that the most important variables in determining ascidian distribution are temperature (Namaguchi et al. 1997), salinity (Vazquez and Young 2000), competition (Lambert 2001), predation (Castilla et al. 2002), light (Tsuda et al. 2003) and hydrodynamics (Holloway and Connell 2002). Besides, the life history traits of ascidians also ensuing the settlement pattern in its habitat. Huus (1927) has shown that few ascidians larvae could be expected to survive oceanic transport over distances greater than 500 km. Olson (1983) reported that all colonial species, embryos are brooded internally and tailed larvae, with free swimming time of little as ten minutes. Similar observation was made by Polk (1962) on *Botryllus schlosseri* in a dock at Ostend, where larval settlement was dense near the parent stocks but sparse only 1 km away. In the present study, maximum numbers of ascidians were recorded from Mandapam coast located at Gulf of Mannar, the east coast of India. It could be correlated with the plenty of suitable substrates available here including natural as well as man-made structures which might have assisted the settlement of the ascidians. Kott (2002), Abdul Jaffar Ali (2004) and Tamilselvi et al. (2011) reported that the provisions of maritime and other installations associated with commercial fishing harbour, facilitates the settlement of ascidians species.

Furthermore these stations are located very close to the Gulf of Mannar, a hot spot area for rich diversity as this biosphere reserve is known to harbour varieties of marine flora and fauna because of its biological and ecological uniqueness.

In contrary, very poor number of ascidians recorded in the stations such as Kayalpattinam, Veerapandianpattinam and Muthunagar beach could be due to the instability of sandy substratum which prevents the attachment of larva. The present observation is confirmed with the reports of Young (1989) who reported that habitat stability is an important criterion for survivorship of an organism and sand movements adversely affects the settlement of its larvae. This survey revealed that substratum is a significant ecological factor which influences the occurrence and distribution of ascidians in the marine environment.

The results showed that distribution of colonial ascidians were found to be more in majority of the stations; whereas, the simple ascidians were restricted to harbour and nearby stations only. Colonial ascidians always colonized nearby the parent colony as they exhibits a kind of asexual reproduction (budding) and brood pouch to retain its larvae. The released short lived larvae facilitate the dispersal of the species over short distances and settled close to the parental colony (Yund and Stires 2002; Tamilselvi et al. 2011). Some ascidians undergo frequent fission, but the resulting colonies remain close to the parents (Bak et al. 1981; Ryland et al. 1984; Stocker 1991). Osman and Whitlatch (1998) noted that colonial species occupy space rapidly and inhibits the recruitment and settlement of other organisms. This could be corroborated with the report of Tsurumi and Reiswig (1997), the fragmented colony reattach to appropriate substratum and are transported to other areas by water currents, tides, and waves. Besides, colonial ascidians were more successful in acquiring space than solitary ascidians with response to increased disturbance magnitude, frequency and community age.

The permanence and dominance nature of the two simple ascidians, *Phallusia nigra* and *P. arabica* in Thoothukudi port indicates the introduced range, survival and successful settlements of these species. Obviously this station is being protected within the two parallel arms limiting the water renewal which in turn turbidity made this area a transformed environment. Provisions of maritime and other permanent structures provide additional habitats for settlement of these species throughout year in this station. Naranjo et al. (1996) recorded the following species *Phallusia mammillata*, *Synoicum argus*, *Styela plicata*, *Clavelina lepadiformis* and *Diplosoma spongiforme* were also frequent seen in transformed environments and less stressed habitats in the bay of Southern Spain.

Even though the simple ascidian *H. momus* was recorded as dominant one among the megafaunal diversity at Thoothukudi harbour in the last decade, but it was noted as rare in the present collection as only one specimen was collected here. The nearby stations Kayalpattinam and Veerapandianpattinam have been the inhabitant of the *H. momus* and collected from 5 m depth to 27 m depth. This species can be regarded as regressive species since the regressive species cannot tolerate in the transformed environment. Simple ascidians belong to the family Pyuridae was common in majority of the stations and were collected from the shallow littoral zone also. *Microcosmus* species have the ability to occupy extensive areas of hard substrata in littoral zones particularly bays and harbours. According to Naranjo et al. (1996), transgressive species are dominant in harbour areas and nearby zones with highly transformed substrates, low rate of water renewal and excess silting and suspended matters. Turon et al. (2007) noticed the dense, mono-specific crusts of *M. squamiger* that out competeed native species in shallow water communities and commonly found in several areas of the Western Mediterranean and Atlantic.

Many ascidians are highly invasive and can spread rapidly to new habitats (Lambert 2002; Rocha and Kremer 2005) and occupied unusually large areas of seafloor and spread rapidly throughout the New England region was reported by Valentine et al. (2007). From our collection, *Herdmania momus*, non-native to Indian coast is commonly distributed and available in plenty at Thoothukudi harbour area between 2001 and 2008; and sporadic distribution of the same species from 2009 to 2011 and afterwards could have been collected from the nearby fish landing centers Kayalpattinum and Veerapandianpattinum, approximately 35 km and 45 km away from the harbour area respectively. According to Naranjo et al. (1996), the regressive species cannot tolerate in the transformed environment particularly in the harbour area where minimal water replenishment, excess silting and suspended matter are experienced. The climate change due to global warming and polluted environment might have caused unfavourable environment to *H. momus*, which in turn, forced this ascidian to migrate nearby non-transformed areas of the aforesaid stations. Shenker and Loya (2008) suggested that this phenomenon is the nature of non-indigenous ascidian crossing its initial stages of invasion or a species that is continuing to spread. Similarly, for the past two decade, *Ecteinascidia venui* is collected from Thoothukudi harbour area only and is not recorded elsewhere. Its distribution in Mandapam is notable and to be investigated.

In contrast to fear all over the world about the non-native species, ascidians play a positive significant role in megabenthic community of marine ecosystem by its dominance and a key-stone species. It also provides a habitat for a number of organisms inside and outside of the body illustrating commensalism or symbiotic relationship with others was reported by Abdul Jaffar Ali (2004) and Tamilselvi et al. (2015) in Indian coastal water. In the present study, several species such as *Trididemnum clinids, Diplosoma swamiensis* and *Lissoclinum bistratum* symbiotically associated with prochloron species. From this, it is clear that ascidians play an important role as a member in maintaining the structure and functioning of marine ecosystem. Tamilselvi et al. (2015) reported that *Herdmania momus* is a model organism for commensal like faunal association, as well as an indicator species for changed environmental conditions in marine community.

9.7 Conclusion

India is one of the richest countries of the world in terms of biological diversity due to its biogeographic position, varied climatic conditions, and bestowed with different types of habitats. Diversity is the central dogma for all other research oriented areas such as pharmacology, aquaculture, immunology, biotechnology, embryology, etc. Recently, scientists have realized the significance of ascidians as they are the only organisms, next to frog showing retrogressive metamorphism and hence moving towards ascidians around the world. Besides, next to sponges, certain species of this group have clinical value, because of their static life forced them to lead successfully in the stressed environment by synthesizing potential bioactive compounds as chemical defense. Ever increasing invasiveness of ascidians warrents the continuous monitoring of marine ecosystem. Ascidians being key stone species in the benthic megafaunal community, the periodic assessment of this group is indispensible. This content will definitely very helpful for researchers, coastal planners, port authorities, coastal thermal plants and atomic power plants for proper management of this ecologically significant ascidians.

Chapter 10
Scope for Future Research in Ascidians

- Recognizing the importance of ascidians for their own intrinsic ecological, commercial and environmental values, more uncharted areas have to be explored for new species.
- Significant effect of supplementary feeds prepared from the tunicates, *Didemnum psammathodes* and *Herdmania momus* (non-native to India), on growth and reproductive performance of ornamental fish, black molly, *Poecilia sphenops* have opened a new avenue to introduce more tunicates as non-conventional resource of food for culturable species. Further this would minimize the over exploitation of existing commercially valuable species.
- Since some ascidians are considered as potential source of bioactive compounds, the diseases of fishes and other organisms can be checked/minimized by providing feeds supplemented with tunicates to increase the yield.
- Sedentary mode life, hermaphrodite gonads and short larval life of ascidians are positive factors to make them breed in controlled water. Hence mass culture practice of this economically valued species could be possible by applying captive breeding techniques.
- Some species of tunicates are identified as bioindicator of environmental degradation, Such species can be used to monitor the environment.
- Presence of specialized pigment tunichromes for iron absorption, as well as, calcium in the form of spicules, ascidians could possibly be used in some of food to treat commonly prevalent diseases such as anaemia, osteoporosis etc.,
- Accumulation of large quantity of vanadium by certain species of ascidians (eg. *Phallusia nigra*) signals the screening of ascidians for the preparation of biofertilizers.
- Tunicate is the only animal known to produce cellulose equivalent. The test of ascidians is made up of tunicin, similar to the cellulose of plants. Biofuels can be prepared from the tunic of ascidians by adopting biotechnological tool.

© Springer International Publishing Switzerland 2016
H.A. Jaffar Ali and M. Tamilselvi, *Ascidians in Coastal Water*,
DOI 10.1007/978-3-319-29118-5_10

- Since this group of animal contains essential aminoacids, high quality proteins and omega-3-fattyacid, this animal can be recommended to prepare value added products such as soup, wafers, pickles, jam or even as ingredients.
- Lack of ascidian taxonomists and limitations in routine morphology based species identification signal the need for a new approach to taxon recognition. DNA barcoding based on the use of a short DNA sequence have been proposed as a rapid and cost effective molecular technique to identify species. The idea of using nucleotide sequences as barcode for ascidian species identification would be of supportive tool to solve the ambiquity in confirming the species.

Bibliography

Abbott, D. P., Newberry, A. T., & Morris, K. M. (1997). *Reef and Shore Fauna of Hawaii. 6B: Ascidians* (*Urochordata*). Honolulu: Bishop Museum Press.

Abdul Jaffar Ali, H. (2004). Comparative study on ecology of *Phallusia nigra* Savigny 1816 from Tuticorin (southeast coast) and Vizhinjam (southwest coast). *Dissertation.* India: Manonmaniam Sundaranar University.

Abdul Jaffar Ali, H. A., & Sivakumar, V. (2007). Occurrence and distribution of ascidians in Vizhinjam Bay (south west coast of India). *Journal of Experimental Marine Biology and Ecology, 342*, 189–190.

Abdul Jaffar Ali, H. A., Sivakumar, V., & Tamilselvi, M. (2010). New records of colonial ascidians from south west coast off India. *Middle East Journal of Scientific Research, 5*(5), 366–373.

Abdul Jaffar Ali, H., Sivakumar, V., & Tamilselvi, M. (2009). Distribution of alien and cryptogenic ascidians along the southern coasts of Indian peninsula. *World Journal of Fish and Marine Sciences, 1*(4), 305–312.

Abdul Jaffar Ali, H., Tamilselvi, M., & Sivakumar, V. (2014). Non-indigenous ascidiansin V. O. Chidambaram port, Thoothukudi, India. *Indian Journal of Geo-Marine Sciences* (in press).

Bak, R. P. M., Sybesma, J., & Van Duyl, F. C. (1981). The ecology of the tropical compound ascidian *Trididemnum solidum*. II. Abundance, growth and survival. *Marine Ecological Progressive Series, 6*, 43–52.

Berrill, N. J. (1950). *The Tunicata with an account of the British Species* (Vol. 133, pp. 1–354). London: Royal Society of Publishers.

Bingham, B. L., & Young, C. M. (1991). Larval behavior of the ascidian *Ecteinascidia turbinata* Herdman; an in situ experimental study of the effects of swimming on dispersal. *Journal of Experimental Marine Biology and Ecology, 145*, 189–204.

Bouchet, J. M. (1962). Etude bionomique d'une fraction de chenal du basin d'Arcachon (chenal du Courbey). *Bulletin de l'Institut Oceanographique de Monaco, 1252*, 1–16.

Carlton, J. T. (1996a). Pattern, process and prediction in marine invasion ecology. *Biology of Conservation, 78*, 97–106.

Carlton, J. T. (1996b). Biological invasions and cryptogenic species. *Ecology, 77*, 1653–1655.

Castilla, J. C., Collins, A. G., Meyer, C. P., Guinez, R., & Lindberg, D. R. (2002). Recent introduction of the dominant tunicate, *Pyura praeputialis* (Urochordata, Pyuridae) to Antofagasta, Chile. *Molecular Ecology, 11*, 1579–1584.

Convention on Biological Diversity. (2002). *A press release on International Day for Biological Diversity.* From http://www.cbd.int/ibd/22/

Das, S. M. (1936). *Herdmania* (*The monascidian of the Indian seas*). Indian Zoological Memoirs, 5. Lucknow.

Das, S. M. (1938). A case of commensalism between a lamellibranch and a monascidian. *Current Science, 7*(3), 114–115.

© Springer International Publishing Switzerland 2016
H.A. Jaffar Ali and M. Tamilselvi, *Ascidians in Coastal Water*,
DOI 10.1007/978-3-319-29118-5

Das, S. M. (1945). On a collection of monascidians from Madras. *Journal of Royal Asiatic Society of Bengal Science, 11*(1), 6–7.

Das, S. M. (1957). *Herdmania*. Allahabad: Indian Press Pub. Pvt. Ltd.

Dybern, B. I. (1963). Biotope choice in *Ciona intestinalis* (L.). Influence of light. *Zoologiska Bidrag Fran Uppsala, 35*, 171–199.

Garstang, W. (1896). Outlines of a new classification of the Tunicata. *Reports of the British Association, 1895*, 718–719.

Garstang, W. (1928). The morphology of the Tunicata, and its bearings on the phylogeny of the Chordata. *Quarterly Journal of Microscopical Science, 72*, 51–187.

Goodbody, I. (1974). The physiology of ascidians. *Advances in Marine Biology, 12*, 1–149.

Goodbody, I. (2003). The Ascidian fauna of Port Royal, Jamaica in Harbour and mangrove dwelling species. *Bulletin of Marine Science, 73*, 457–476.

Griffiths, D. J., & Thinh, L. V. (1983). Transfer of photosynthetically fixed carbon between the prokaryotic green alga Prochloron and its ascidian host. *Australian Journal of Marine and Freshwater Research, 34*, 431–440.

Harent, H. (1951). Les Tuniciers comestibles. *Attidel Congresso Internazianale d' Igienee di Medieina Mediterranea Palermo, 118*, 1–3.

Hayward, P. J., & Ryland, J. S. (1990). The marine fauna of the British Isles and north-west Europe. Oxford: Oxford University Press.

Herdman, W. (1882). Report on the Tunicata collected during the voyage of H.M.S. "Challenger" during the years 1873–1876. Part I. Ascidiae simplices. *ZChallengeroology of the 'Challenger' Expedition, 6*, 1–296.

Hirose, E., Oka, A. T., & Akahori, M. (2005). Sexual reproduction of the photosymbiotic ascidian Diplosoma virens in the Ryukyu Archipelago, Japan: vertical transmission, seasonal change, and possible impact of parasitic copepods. *Marine Biology, 146*, 677–682.

Holloway, M. G., & Connell, S. D. (2002). Why do floating structures create novel habitats for subtidal epibiota? *Marine Ecological Progressive Series, 235*, 43–52.

Huus, J. (1927). Uber die Ausbreitungshindernisse derMeerestiefen und die geographische Verbreitung derAscidien. *Nyt Magasin for Naturvetenskap, 65*, 153–174.

Huus, J. (1937). Tunicata. Ascidiacea. *Handbord of zoology. Kiikenthal and Krumbach, 5*, 542–672.

John, H. (1986). Marine natural products as lead to new pharmaceutical and agrochemical agents. *Pure and Applied Chemistry, 58*(3), 365–374.

Kott, P. (1963). The ascidians of Australia IV. Aplousobranchia Lahille, Polyclinidae Verrill (continued). *Australian Journal of Marine and Freshwater Research, 14*(1), 70–118.

Kott, P. (1972). The ascidians of South Australia I. Spencer Gulf, St Vincent Gulf and Encounter Bay. *Transactions of the Royal Society of South Australia, 96*(1), 1–52.

Kott, P. (1985). The Australian Ascidiacea. *Part I, Phlebobranchia and Stolidobranchia in Memories of Queensland Museum, 23*, 1–440.

Kott, P. (1990). The Australian Ascidiacea. *Phlebobranchia and Stolidobranchia, supplement. Memoirs of the Queensland Museum, 29*(1), 267–298.

Kott, P. (1992). The Australian Ascidiacea, Pt 3 Aplousobranchia (2). *Memoirs of the Queensland Museum, 32*(2), 377–620.

Kott, P. (1997). Ascidians. In S. Shepherd, & M. Davies (Eds.), Marine Invertebrates of Southern Australia part III. (South Australian Research and Development Institute [Aquatic Sciences with the Flora and Fauna of South Australia Handbooks Committee]: Adelaide). pp. 1107–1280.

Kott, P. (2001). The Australian Ascidiacea. Part IV: Aplousobranchia (3). *Didemnidae in Memoirs of the Queensland Museum., 47*, 1–410.

Kott, P. (2002). A complex didemnid ascidian from Whangamat, New Zealand. *Journal of Marine Biological Association of UK, 82*(4), 625–628.

Kott, P. (2003). New syntheses and new species in the Australian Ascidiacea. *Journal of Natural History, 37*, 1611–1653.

Kott, P. (2008). Ascidiacea (Tunicata) from deep waters of the continental shelf of Western Australia. *Journal of Natural History, 42*(15–16), 1103–1217.

Krishnan, R., Chandran, M. R., & Renganathan, T. K. (1989). On the occurrence of four species of ascidians new to Indian waters. *Geobios new Reports, 8,* 70–74.

Lahille, F. (1886). Sur la classification des Tuniciers. *Comptes Rendus de l'Academie des Sciences Paris, 102,* 1573–1575.

Lambert, G. (2001). A global overview of ascidian introductions and their possible impact on the endemic fauna. In H. Sawada, H. Yokosawa, & C. C. Lambert (Eds.), *The biology of ascidians* (pp. 249–257). Tokyo: Springer-Verlag.

Lambert, C. (2002). Non-indigenous ascidians in tropical waters. *Pacific Science, 56,* 291–298.

Marshall, D. J., & Keough, M. J. (2003). Variation in the dispersal potential of non-feeding invertebrate larvae: the desperate larva hypothesis and larval size. *Marine Ecological Progressive Series, 255,* 145–153.

Meenakshi, V. K. (1997). Biology of a few chosen ascidians. *Dissertation.* India: Manonmaniam Sundaranar University..

Meenakshi, V. K. (1998a). Occurrence of a new ascidian species—*Distaplia nathensis* sp. nov. and two species—*Eusynstyela tincta* (Van Name, 1902), *Phallusia nigra* (Savigny, 1816) new records for Indian waters. *Indian Journal of Marine Science, 27,* 477–479.

Meenakshi, V. K. (1998b). Three species of polyclinid ascidians—new records to Indian waters. *Journal of Marine Biological Association of India, 40*(1&2), 201–205.

Meenakshi, V. K. (2002). Occurrence of a new species of colonial ascidian—*Eudistoma kaverium* sp. nov. and four new records of *Eudistoma* to Indian coastal waters. *Indian Journal of Marine Science, 31*(3), 201–206.

Meenakshi, V. K. (2003). *Marine biodiversity—taxonomy of indian ascidians* (pp. 1–103). New Delhi: Final Technical Report submitted to the Ministry of Environment and Forests.

Meenakshi, V. K. (2004). Conservation strategies and action plan for the prochordates in the Journal of Tamil Nadu Biodiversity Strategy and Action Plan. *Chordate Diversity,* 17–30.

Meenakshi, V. K., & Senthamarai, S. (2004). First report of a simple ascidian—*Phallusia arabica* Savigny, 1816 from Tuticorin coast of India. *Journal of the Marine Biological Association of India, 46*(1), 104–107.

Meenakshi, V. K., & Senthamarai, S. (2006). First report on two species of ascidians to represent the genus *Botryllus* Gaertner, 1774 from Indian water. *Journal of the Marine Biological Association of India, 48*(1), 1–102.

Meenakshi, V. K., & Senthamarai, S. (2007). New records of two species of simple ascidians—*Microcosmus pupa* (Savigny, 1816) and *Microcosmus squamiger,* Hartmeyer and Michaelsen, 1928—from Indian seas. *Journal of the Bombay Natural History Society, 104*(2), 238–240.

Meenakshi, V. K., & Senthamarai, S. (2013). Diversity of ascidians from the Gulf of Mannar. In K. Venkataraman, C. Sivaperuman, & C. Raghunathan (Eds.), *Ecology and conservation of tropical marine faunal communities* (pp. 213–229). Berlin: Springer.

Miller, R. H. (1975). Ascidians from the Indo-West Pacific region in the Zoological Museum, Copenhagen (Tunicata, Ascidiacea). *Steenstrupia, 3*(20), 205–306.

Monniot, F. (1983). Ascidies littorales de Guadeloupe I. Didemnidae. *Bulletin du Museum National d'Histoire Naturelle, 4*(5), 5–49.

Monniot, F., & Monniot, C. (1996). New collections of ascidians from the western Pacific and southeastern Asia. *Micronesia, 29,* 133–279.

Monniot, F., & Monniot, C. (2001). Ascidians from the tropical western Pacific Zoosystema, 23*(2), 201–383.

Namaguchi, T. A., Nishijima, C., Minowa, S. Hashimoto, M., Haraguchi, C., Amemiya, S., et al. (1997). Embryonic thermosensitivity of the ascidian, *Ciona savigny. Zoological Science,* (Tokyo), 14, 511–516.

Naranjo, S. A., Carballo, J. L., & Garcia-Gomez, J. C. (1996). Effects of environmental stress on ascidian populations in Algeciras Bay (Southern Spain). Possible marine bioindicators? *Marine Ecological Progressive Series, 144,* 119–131.

Nishikawa, T., & Tojcioka, T. (1976). Ascidians from the Amami Islands. (Contributions to the Japanese ascidian fauna XXVIII). *Publications of the Seto Marine Biological Laboratory, 22*(6), 377–402.

Oka, A. (1915). Report on the Tunicata in the collection of the Indian Museum. *Memories of Indian Museum, 6*, 1–33.

Oka, A. (1931). Ueber Myxobotrus, eine neue Synascidien-Gattung. *Proceedings of the Imperial Academy, 7*, 238–240.

Olson, R. R. (1983). Ascidian—Prochloron symbiosis: the role of larval photoadaptations in midday larval release and settlement. *Biological Bulletin, 165*(1), 221–240.

Osman, R. W., & Whitlatch, R. B. (1998). Local control of recruitment in an epifaunal community and the consequences to colonization processes. *Hydrobiology, 375*(376), 113–123.

Pardy, R. L., & Lewin, R. A. (1981). Colonial ascidians with prochlorophyte symbionts: evidence for translocation of metabolites from alga to host. *Bulletin of Marine Science, 31*, 817–823.

Perrier, J. O. E. (1898). Note sur la classification des Tuniciers. *Comptes Rendus de l'Academie des Sciences Paris, 126*, 1758–1762.

Polk, P. (1962). Bijdrage tot de kenntnis der mariene fauna van de belgische kust. II. Waarnemingen aangande het voorkomen en de voortplanting van Botryllus schlosseri (Pallas, 1766). *Natuurw Tidjschr, 44*, 21–28.

Renganathan, T. K. (1981). On the occurrence of colonial ascidian, *Didemnum psammathodes* (Sluiter, 1985) from India. *Current Science, 50*(20), 922.

Renganathan, T. K. (1982a). On the occurrence of colonial ascidian *Lissoclinum fragile* (Van Name, 1902) from India. *Current Science, 51*(3), 149.

Renganathan, T. K. (1982b). New record of a genus of colonial ascidian from India. *Current Science, 51*(5), 253–254.

Renganathan, T. K. (1982c). Observation on the presence of embryos and larvae in the atrial cavity of *Eudistoma* sp. of Tuticorin Coast. *Life Science Advances, 1*(2), 191–192.

Renganathan, T. K. (1983a). First record of a simple ascidian *Microcosmus curvus* Tokioka, 1954 from Indian waters. *Current Science, 52*(19), 929–930.

Renganathan, T. K. (1983b). *Pyura* Molina, 1782 a recorded genus of simple ascidian from India. *Geobios New Reports, 2*, 57–58.

Renganathan, T. K. (1983c). *Perophora formosana* Oka, 1931. (Ascidiacea: Perophoridae)—a new record for the Indian waters. *Geobios New Reports, 2*, 78–79.

Renganathan, T. K. (1983d). Breeding season of a colonial ascidian, *Didemnum psammathodes* (Sluiter, 1875) of Tuticorin Coast, India. *Journal of Biological Research, 3*(1), 54–56.

Renganathan, T. K. (1983e). A note on gonadial variation in a colonial ascidian, *Eudistoma* sp. of Tuticorin Coast. *Geobios New Reports, 2*, 78–79.

Renganathan, T. K. (1984). *Ecteinascidia garstangi*, Sluiter 1898—a colonial ascidian not hitherto been recorded from India. *Geobios New Reports, 3*, 54–55.

Renganathan, T. K. (1984a). New record and redescription of a rare colonial ascidian *Eudistoma viride* Tokioka 1955 from the Indian waters. *Geobios New Reports, 3*, 49–51.

Renganathan, T. K. (1984b). *Aplidium multiplicatum*, Sluiter, 1909—a new record for the Indian waters. *Geobios New Reports, 3*, 155–156.

Renganathan, T. K. (1985). On the occurrence of a colonial ascidian, *Symplegma brakenhielmi* Michaelsen, 1904 from Tuticorin coast of India. *Geobios New Report, 4*, 75–77.

Renganathan, T. K. (1986). Studies on the ascidians of South India. *Dissertation*. India: Kamaraj University.

Renganathan, T. K., & Krishnaswamy, S. (1985). Some colonial ascidians from Indian waters. *Indian Journal of Marine Science, 14*, 38–41.

Rocha, R. M., & Kremer, L. P. (2005). Introduced ascidians in Paranagua Bay, Parana, Southern Brazil. *Revista Brasileira de Zoologia, 22*(4), 1170–1184.

Rodrigues, S. A., Rocha, R. M., & Lotufo, T. M. C. (1998). Guia ilustrado para identificação das ascídias do Estado de Sao Paulo. Sao Paulo: Parma

Ryland, J. S., Wiley, R. A., & Muirhead, A. (1984). Ecology and colonial dynamics of some Pacific reef flat Didemnidae (Ascidiacea). *Zoological Journal of the Linnean Society, 80*, 261–282.

Sanders, H. L. (1960). Benthic studies in Buzzards Bay. III. The structure of the soft-bottom community. *Limnology and Oceanography, 5*, 138–153.

Sebastian, V. O. (1952). A new species of synascidian from Madras. *Current Science, 21*, 316–317.

Sebastian, V. O. (1954). On *Polyclinum indicum*, a new Synascidian from the Madras Coast of India. *Journal of the Washington Academy of Sciences, 44*(1), 18–24.

Sebastian, V. O. (1955). *Perophora listeri indica* var. nova, a new ascidians from Madras Coast of India. *Zoologischer Anzeiger, 154*(11 & 12), 266–288.

Sebastian, V. O. (1956). *Symplegma viride* Herdman and *Symplegma viride* Stolonica Berril, two unrecorded fouling organisms from Indian seas. *Journal of Timber Dryers and Preservers Association of India, 11*(3), 2–4.

Sebastian, V. O., & Kurian, C. V. (1981). Indian ascidians. Oxford and IBH Publishing Co. New Delhi, pp: 1–144.

Shenker, N., & Loya, Y. (2008). The solitary ascidian *Herdmania momus*: native (Red Sea) versus non-indigenous (Mediterranean) populations. *Biological Invasions, 10*, 1131–1439.

Stocker, L. J. (1991). Effects of size and shape of colony on rates of fusion, growth and mortality in a subtidal invertebrate. *Journal of Experimental Marine Biology and Ecology, 14*, 161–175.

Storer, T. I., & Usinger, R. L. (1965). *General zoology*. Bombay—New Delhi: Tata Mc Graw Hill.

Tamilselvi, M. (2008). Ecological studies on ascidians of Tuticorin coast. *Dissertation, Manonmaniam Sundaranar University, India.*

Tamilselvi, M. (2014). *Feed formulation from exotic ascidians Didemnum psammathodes and Herdmania pallida.* New Delhi: A project report submitted to UGC.

Tamilselvi, M., Abdul Jaffar Ali, H., Sivakumar, V., & Kanagaraj, G. (2015). Consumption assessment qualities of pellet feed from *Herdmania momus*, non-indigenous tunicate from Thoothukudi coastal area. (*Paper presented at the National Conference on Current and Emerging Trends in Biological Science*, Virudhunagar, India).

Tamilselvi, M., Abdul Jaffar Ali, H., & Thilaga, R. D. (2012). Diversity and seasonal variations of class Ascidiacea in Thoothukudi coast, India. *International Journal of Environmental Sciences, 3*(3), 1097–1115.

Tamilselvi, M., Sivakumar, V., Abdul Jaffar Ali, H., & Thilaga, R. D. (2011). Distribution of alien tunicates (Ascidians) in Tuticorin coast, India. *World Journal of Zoology, 6*(2), 164–172.

Tokioka, T. (1954). Invertebrate fauna of the intertidal zone of the Tokara Islands, VII Ascidians (Contributions to Japanese ascidian fauna VII). *Publications of the Seto Marine Biological Laboratory, 3*(3), 239–264.

Tokioka, T. (1967). Pacific Tunicata of the United States National Museum. *Bulletin of US National Museum, 251*, 1–242.

Tsuda, M., Kusakabe, T., Iwamoto, H., Horie, T., Nakashima, Y., Nakagawa, M., et al. (2003). Origin of the vertebrate visual cycle: II. Visual cycle proteins are localized in whole brain including photoreceptor cells of a primitive chordate. *Vision Research, 43*, 3045–3053.

Tsurumi, M., & Reiswig, H. M. (1997). Sexual versus asexual reproduction in an oviparous rope-form sponge. *Aplysina cauliformis* (Porifera:Verongida). *Invertebrate Reproduction and Development, 32*, 1–9.

Turon, X., Nishikawa, T., & Rius, M. (2007). Spread of *Microcosmus squamiger* (Ascidiacea: Pyuridae) in Mediterranean Sea and adjacent waters. *Journal of Experimental Marine Biology and Ecology, 342*, 185–188.

Valentine, P. C., Collie, J. S., Reid, R. N., Asch, R. G., Guida, V. G., & Blackwood, D. S. (2007). The occurrence of the colonial ascidian *Didemnum* sp. on Georges bank gravel habitat-Ecological observations and potential effects on ground fish and scallop fisheries. *Journal of Experimental Marine Biology and Ecology,342*, 179–181.

Van Name, W. G. (1945). The North and South American ascidians. *Bulletin of the American Museum of Natural History, 84,* 1–476.

Vazquez, E., & Young, C. M. (2000). Effects of low salinity on metamorphosis in estuarine colonial ascidians. *Invertebrate Biology, 119,* 433–444.

WWF. (2009). *Silent invasion—the spread of marine invasive species via ships' ballast water* (p. 22). Gland: WWF International.

Young, C. M. (1989). Distribution and dynamics of an intertidal ascidian pseudopopulation. *Bulletin of Marine Sciences, 45*(2), 288–303.

Yund, P. O., & Stires, A. (2002). Spatial variation in population dynamics in a colonial ascidian (*Botryllus schlosseri*). *Marine Biology, 141,* 955–963.

Species Index

© Springer International Publishing Switzerland 2016
H.A. Jaffar Ali and M. Tamilselvi, *Ascidians in Coastal Water*,
DOI 10.1007/978-3-319-29118-5

Subject Index

A
Adhesive papillae, 11, 124
Alien, 24
Amphipod, 28
Antibacterial, 37
Antifouling, 34, 38
Antifungal, 35
Antitumour, 34, 37, 38
Aplousobranchia, 13, 15, 34, 38, 73, 74, 77,
    79, 81, 83, 85, 87, 89, 90, 92, 94, 97, 98,
    101, 103, 105–107, 110, 113, 115, 117,
    118, 120, 124, 126, 128
Appendicularia, 14
Arabian Sea, 43
Ascidiacea, 3, 13, 14, 51, 52, 54–60, 62,
    64–71, 73, 74, 77, 79, 81, 83, 85, 87, 89,
    90, 92, 94, 97, 98, 101, 103, 105–107, 110,
    113, 115, 117, 118, 120, 124, 126, 128, 130
Ascidiologist, 27
Associated fauna, 27
Associated flora, 29
Association, 22, 28, 140
Atrial cavity, 10, 79, 83
Atrial siphon, 8, 9, 15, 16, 29, 54, 56–59, 70,
    74, 77, 79, 81, 83, 87, 110, 113, 114, 117

B
Bacteria, 29, 37
Ballast water, 23, 24, 43
Barbed spines, 27, 73
Barges, 22, 38, 47, 48
Bay of Bengal, 43
Benthic organism, 25, 33
Biofouler, 000
Bio-indicator, 25, 26, 143
Bivalve, 28, 47
Black ascidian, 26, 58
Boulders, 22, 38, 45, 48

Branchial siphon, 15, 16, 29, 37, 52, 57, 59,
    67–71, 73, 74, 83, 113, 114, 117, 123, 127
Breeding, 21, 23, 24, 143
Bryozoans, 24, 34, 38
Budding, 10, 15, 16, 21, 39, 139

C
Cadmium, 26
Chordata, 39
Clliary, 4, 7
Cloacal aperture, 5, 50, 65, 76, 83, 94, 103,
    106, 113, 118, 128
Cobalt, 26
Colonial ascidian, 5, 10, 14, 15, 20–22, 28, 29,
    32, 37–40, 48, 50, 52, 79, 81, 90, 136, 139
Colony, 5, 10, 17, 21, 29, 50, 61, 64, 65, 76,
    77, 79, 83, 87, 89, 90, 93, 98, 101, 106,
    114, 116, 119, 120, 124, 126, 128, 130
Commensalism
Concrete submerged blocks, 22
Copper, 26
Corals, 22
Cryptogenic, 24, 51
Cytotoxic, 37, 38

D
Deep sea, 19, 20, 45, 48
Diazonamide A, 34
Didemnid, 29
Didemnin-B, 34, 38
Dislodging, 48
Dispersal, 10, 20, 23, 39, 138, 139
Dissolved oxygen, 22
Distribution, 19–22, 29, 31, 44, 51, 52, 54,
    57–60, 64, 65, 67, 68, 70, 71, 73, 74,
    77–79, 81, 83, 86, 87, 89, 90, 93, 94, 98,
    101, 103, 106, 110, 113, 116, 117, 119,
    120, 124–130, 138–140

© Springer International Publishing Switzerland 2016
H.A. Jaffar Ali and M. Tamilselvi, *Ascidians in Coastal Water*,
DOI 10.1007/978-3-319-29118-5

Sentinel, 25
Sessile, 3, 4, 10, 20–23, 25, 33, 39, 73, 94, 130
Shallow water, 19, 45, 48, 52, 54, 64, 67, 78,
    120, 129, 140
Shellfish reef, 45
Snorkeling, 45
Solitary ascidian, 5, 18, 20, 22, 29, 32, 37, 38,
    50, 57, 58, 70, 71, 136, 139
Sorberacea, 14
Spermduct, 16
Spicules, 5, 16, 21, 51, 71, 113, 114, 117, 119,
    120, 123, 128, 130, 143
Sponges, 5, 24, 34, 38, 79, 141
Stigmata, 7, 11, 15–18, 50, 51, 56–61, 64, 66,
    70, 76, 79, 81, 83, 87, 93, 96, 98, 99, 101,
    104, 106, 110, 111, 114, 120, 123, 124,
    126, 128
Stolidobranchia, 13, 38, 59, 60, 62, 64–71, 73
Substratum, 5, 10, 11, 19, 22, 23, 52, 57, 60,
    70, 138, 139
Super organism, 25, 27
Supplementary feed, 31, 32, 143
Symbiont, 29, 115, 117
Synascidians, 5, 50, 136

T

Tadpole, 4, 5, 10, 11, 17, 20, 22, 130
Temperature, 21–23, 29, 138
Test, 3–5, 8, 10, 11, 14, 21, 26, 27, 37, 39, 51,
    54, 57–60, 64, 65, 67, 70, 71, 73, 74, 77,
    79, 81, 83, 87, 89, 94, 96–98, 101, 103,
    106, 110, 113, 114, 117, 119, 120, 125, 128
Thaliacea, 13, 14
Thorax, 10, 14–17, 51, 77, 79, 81, 83, 87, 90,
    93, 94, 98, 103, 110, 117, 128
Transgressive, 25, 26, 140
Tunic, 3, 21, 22, 56, 69, 73, 143
Tunicate, 3–5, 8, 9, 13, 14, 19–28, 31, 32, 34,
    37–39, 143
Tunichrome, 9, 25, 143
Tunicin, 3, 143
Turbidity, 22, 139

U

Urochordata, 3, 4, 31, 39

V

Vanadium, 9, 21, 26, 33, 143
Vanodocytes, 4, 8, 9, 25, 26
Vitilevuamide, 35

W

Walking access, 49
Water current, 7, 10, 22, 139

Z

Zinc, 26
Zooids, 5, 10, 15–17, 21, 40, 50, 52, 54, 60, 61,
    64–66, 77, 81, 83, 87, 89, 90, 94, 100, 103,
    106, 110, 113, 116, 119, 123–126, 128, 130